Learn

Eureka Math®

Grade 3
Modules 3 & 4

TEKS EDITION

Great Minds® is the creator of *Eureka Math®*, *Wit & Wisdom®*, *Alexandria Plan™*, and *PhD Science®*.

Published by Great Minds PBC
greatminds.org

Printed in Mexico

1 2 3 4 5 6 7 8 9 10 QMX 27 26 25 24 23

ISBN 978-1-63642-860-4

Learn • Practice • Succeed

Eureka Math® student materials for *A Story of Units®* (K–5) are available in the *Learn, Practice, Succeed* trio. This series supports differentiation and remediation while keeping student materials organized and accessible. Educators will find that the *Learn, Practice,* and *Succeed* series also offers coherent—and therefore, more effective—resources for Response to Intervention (RTI), extra practice, and summer learning.

Learn

Eureka Math Learn serves as a student's in-class companion where they show their thinking, share what they know, and watch their knowledge build every day. *Learn* assembles the daily classwork—Application Problems, Exit Tickets, Problem Sets, templates—in an easily stored and navigated volume.

Practice

Each *Eureka Math* lesson begins with a series of energetic, joyous fluency activities, including those found in *Eureka Math Practice.* Students who are fluent in their math facts can master more material more deeply. With *Practice,* students build competence in newly acquired skills and reinforce previous learning in preparation for the next lesson.

Together, *Learn* and *Practice* provide all the print materials students will use for their core math instruction.

Succeed

Eureka Math Succeed enables students to work individually toward mastery. These additional problem sets align lesson by lesson with classroom instruction, making them ideal for use as homework or extra practice. Each problem set is accompanied by a Homework Helper, a set of worked examples that illustrate how to solve similar problems.

Teachers and tutors can use *Succeed* books from prior grade levels as curriculum-consistent tools for filling gaps in foundational knowledge. Students will thrive and progress more quickly as familiar models facilitate connections to their current grade-level content.

Students, families, and educators:

Thank you for being part of the *Eureka Math*® community, where we celebrate the joy, wonder, and thrill of mathematics.

In the *Eureka Math* classroom, new learning is activated through rich experiences and dialogue. The *Learn* book puts in each student's hands the prompts and problem sequences they need to express and consolidate their learning in class.

What is in the Learn book?

Application Problems: Problem solving in a real-world context is a daily part of *Eureka Math.* Students build confidence and perseverance as they apply their knowledge in new and varied situations. The curriculum encourages students to use the RDW process—Read the problem, Draw to make sense of the problem, and Write an equation and a solution. Teachers facilitate as students share their work and explain their solution strategies to one another.

Problem Sets: A carefully sequenced Problem Set provides an in-class opportunity for independent work, with multiple entry points for differentiation. Teachers can use the Preparation and Customization process to select "Must Do" problems for each student. Some students will complete more problems than others; what is important is that all students have a 10-minute period to immediately exercise what they've learned, with light support from their teacher.

Students bring the Problem Set with them to the culminating point of each lesson: the Student Debrief. Here, students reflect with their peers and their teacher, articulating and consolidating what they wondered, noticed, and learned that day.

Exit Tickets: Students show their teacher what they know through their work on the daily Exit Ticket. This check for understanding provides the teacher with valuable real-time evidence of the efficacy of that day's instruction, giving critical insight into where to focus next.

Templates: From time to time, the Application Problem, Problem Set, or other classroom activity requires that students have their own copy of a picture, reusable model, or data set. Each of these templates is provided with the first lesson that requires it.

Where can I learn more about Eureka Math *resources?*

The Great Minds® team is committed to supporting students, families, and educators with an ever-growing library of resources, available at gm.greatminds.org/math-for-texas. The website also offers inspiring stories of success in the *Eureka Math* community. Share your insights and accomplishments with fellow users by becoming a *Eureka Math* Champion.

Best wishes for a year filled with aha moments!

Jill Diniz

Jill Diniz
Director of Mathematics
Great Minds

The Read–Draw–Write Process

The *Eureka Math* curriculum supports students as they problem-solve by using a simple, repeatable process introduced by the teacher. The Read–Draw–Write (RDW) process calls for students to

1. Read the problem.
2. Draw and label.
3. Write an equation.
4. Write a word sentence (statement).

Educators are encouraged to scaffold the process by interjecting questions such as

- What do you see?
- Can you draw something?
- What conclusions can you make from your drawing?

The more students participate in reasoning through problems with this systematic, open approach, the more they internalize the thought process and apply it instinctively for years to come.

Contents

Module 3: Multiplication and Division with Units of 0, 1, 6–9, and Multiples of 10

Module 4: Multiplication and Area

Topic A: Concepts of Area Measurement

Topic B: Arithmetic Properties Using Area Models

Topic C: Applications of Area Using Side Lengths of Figures

Grade 3
Module 3

Ken has a set of 975 building blocks, and his brother, Robert has a set of 1,027 building blocks. How many more building blocks does Robert have compared to Ken? Use strip diagrams to show your work.

Read **Draw** **Write**

Name ______________________________ Date ______________

Use strips of paper to solve. Sketch and label your work. Write an equation that matches the problem. The first one is done for you.

A blue paper strip is 3 times as long as a red paper strip. If the red paper strip is 1 unit long, how long is the blue paper strip?

Red [1]

Blue [| |]

?

$3 \times 1 = 3$

The blue strip is 3 units long.

1. A blue paper strip is 1 unit long. A red paper strip is 4 times as long as the blue paper strip. How many units long is the red paper strip?

2. A red paper strip is twice as long as a blue paper strip. The blue paper strip is 1 unit long. How long is the red paper strip?

3. A red paper strip is 5 units long. The red paper strip is 5 times as long as the yellow paper strip. How long is the yellow paper strip?

4. A red paper strip is 1 unit long. A blue paper strip is 3 times as long as the red paper strip. A yellow paper strip is 2 times as long as the blue paper strip. How long is the yellow strip? How many times longer is the yellow paper strip than the red paper strip?

5. John walks 1 block home from school. Betty walks 4 times as far as John. How many blocks does Betty walk from home to school?

6. A blue paper strip is 3 times as long as a red paper strip. The red paper strip is 1 unit long. How long is the blue paper strip?

   The sketch shows Hayley's work. Explain her mistake.

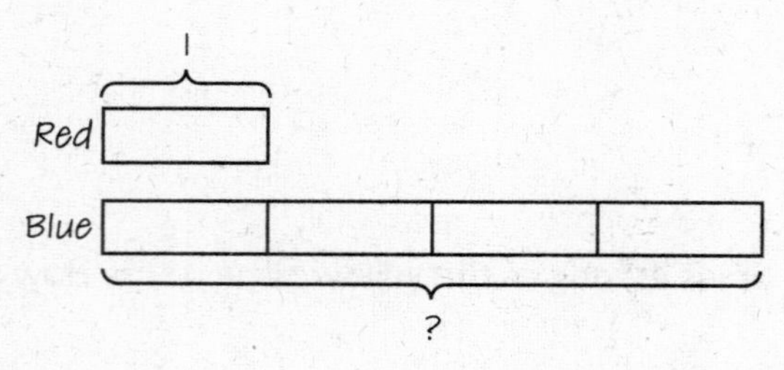

$1 \times 4 = 4$ The blue paper strip is 4 units long.

Name ______________________________ Date ______________

Use strips of red and blue paper to show these comparison. Sketch and label your work. Write an equation, and answer the question.

1. A red paper strip is 4 times as long as a blue paper strip. The blue paper strip is 1 unit long. How long is the red paper strip?

2. A yellow strip of paper is 3 units long. The yellow strip is 3 times as long as a red paper strip. How long is the red paper strip?

On Monday, Rosie was able to jump rope for 1 minute without stopping. By Friday, she could jump rope 5 times longer than she could on Monday. How many minutes was Rosie able to jump rope without stopping on Friday?

Read **Draw** **Write**

Name ______________________________ Date ______________

Solve. Show your work with strip diagrams and an equation.

1. A blue paper strip is 8 centimeters long. A red paper strip is twice as long as the blue paper strip. How long is the red paper strip?

2. The walk from Emma's house to the book store is 3 times as long as a walk from her house to the library. The walk to the library is 5 blocks long. How long is the walk to the book store?

3. Beth bought a ring for $8. Kristen bought a ring that cost 4 times as much as Beth's ring. How much did Kristen's ring cost?

4. An elm tree is 50 feet tall. It is 5 times as tall as the top of the basketball hoop. How tall is the basketball hoop?

5. Three friends were hunting for fossils. Michael found 3 fossils, and Isaac found 4 times as many fossils as Michael found. Garrett found 2 times as many fossils as Isaac. How many fossils did the boys find in all?

6. A wrap costs 2 times as much as a bagel. A salad costs 3 times as much as the wrap. If the salad cost $12, how much does the bagel cost? How much does the wrap cost?

Name ______________________________ Date ________________

Solve using strip diagrams, and an equation.

1. A blue paper strip is 2 centimeters long. A red paper strip is 3 times as long as the blue paper strip. How long is the red paper strip?

2. A yellow paper strip is 45 inches long. It is 5 times as long as a blue paper strip. How long is the blue paper strip?

Ellen knows she can walk a lot faster than her little sister, Diane. For every 3 times Ellen can walk around the track, her sister can only go one time around the track. If they continue at this speed and Ellen walks 15 laps, how many laps will Diane walk?

Read **Draw** **Write**

Name ______________________________ Date ______________

The tables show what happens when a magician's box uses a rule to change numbers. Complete the tables using the rule that is given for each box.

1. The number that comes out of the box is 4 times as much as the number that went into the box.

In	Out
1	
2	
	12
4	
5	
	40

2. The number that comes out of the box is 10 times as much as the number that went into the box.

In	Out
1	
2	
	30
4	
5	
	100

3. The number that comes out of the box is 3 times as much as the number that went into the box.

In	1	3			5	
Out			24	21		27

Complete the table. Then, complete the sentence using the words *times as much.*

4.

In	Out
9	18
8	
	10
3	6
1	

Rule:
The number that comes out is ______________________________ as the number that went in.

5.

In	Out
4	
	30
8	40
10	50
	15

Rule:
The number that comes out is __________________________ as the number that went in to the box.

6. A magician put 4 cards into his magic box. 16 cards came out. Use the words *times as much* to compare 16 and 4.

Name ________________________________ Date ________________

1. Iggy and Nora are saving their allowences. Nora saves three times as much as Iggy. Fill in the table to show how much money Iggy and Nora save.

Iggy	Nora
1	
	9
5	
10	
	27

2. The table shows what happens when the magician puts a number in his magic box. Fill in the blank using the words *times as much*.

In	Out
2	4
10	20
8	16
3	6

The number that comes out of the box is ______________________________ as the number that goes into the box.

Name ________________________________ Date ________________

Solve. Show your work by drawing strip diagrams.

1. Two friends like to look for lost pennies on the sidewalk. George found 9 pennies and Carol found 4 times as many pennies as George. How many pennies did Carol find?

2. Bruce rode his bike on Monday and Friday. On Friday, he rode 5 times as far as he did on Monday. If he rode 30 miles on Friday, how far did he ride on Monday?

3. Peter made a collage using 6 different colorful leaves. His sister, Sue, used 42 colorful leaves in her collage. How many times more leaves did Sue use than Peter?

4. It takes 2 balls of yarn to knit a pair of mittens. A shawl takes 3 times as much yarn as the mittens, and a sweater takes 6 times as much yarn as the mittens. If Grandma wants to knit all three garments, how many balls of yarn does she need?

5. Jake made 20 birdhouses. That was 4 times as many birdhouses as Henry made. Leah made 3 times as many birdhouses as Henry. How many birdhouses did Henry make? How many birdhouses did Leah make?

Solve. Show your work by drawing strip diagrams.

1. Joan owes $6 dollars in library book fines. Stan owes the library 4 times as much as Joan. How much does Stan owe in library fines?

2. Ellie practiced the piano on Tuesday and Friday. On Friday, Ellie practiced for 45 minutes, which was 5 times as much as she practiced on Tuesday. How long did she practice on Tuesday?

Geri brings 3 water jugs to her soccer game to share with teammates. Each jug contains 6 liters of water. How many liters of water does Geri bring?

Read **Draw** **Write**

Name ______________________________   Date ______________

1. a. Solve. Shade in the multiplication facts that you already know. Then, shade in the facts for sixes, sevens, eights, and nines that you can solve using the commutative property.

×	1	2	3	4	5	6	7	8	9	10
1		2	3							
2		4		8				16		
3						18				
4					20					
5										50
6		12								
7										
8										
9										
10										

b. Complete the table. Each bag contains 7 apples.

Number of Bags	2		4	5	
Total Number of Apples		21			42

2. Use the array to write two different multiplication sentences.

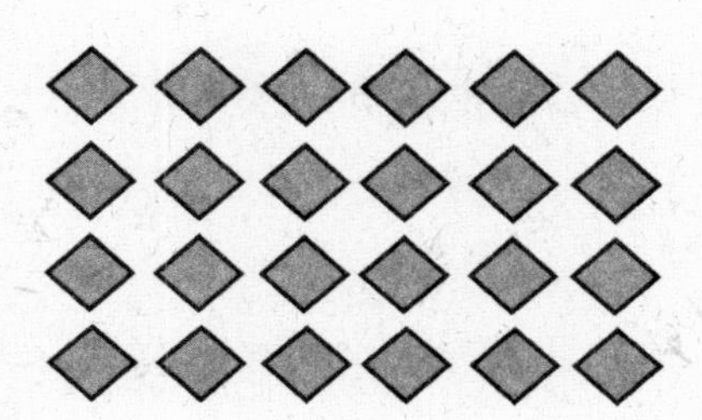

_______ = _______ × _______

_______ = _______ × _______

3. Complete the equations.

a. 2 sevens = ________ twos

= __14__

b. 3 ________ = 6 threes

= ________

c. 10 eights = 8 ________

= ________

d. 4 × ________ = 6 × 4

= ________

e. 8 × 5 = ________ × 8

= ________

f. ________ × 7 = 7 × ________

= __28__

g. 3 × 9 = 10 threes – ________ three

= ________

h. 10 fours – 1 four = ________ × 4

= ________

i. 8 × 4 = 5 fours + ________ fours

= ________

j. ________ fives + 1 five = 6 × 5

= ________

k. 5 threes + 2 threes = _____ × _____

= ________

l. ________ twos + ________ twos = 10 twos

= ________

Name ______________________________ Date ______________

1. Use the array to write two different multiplication facts.

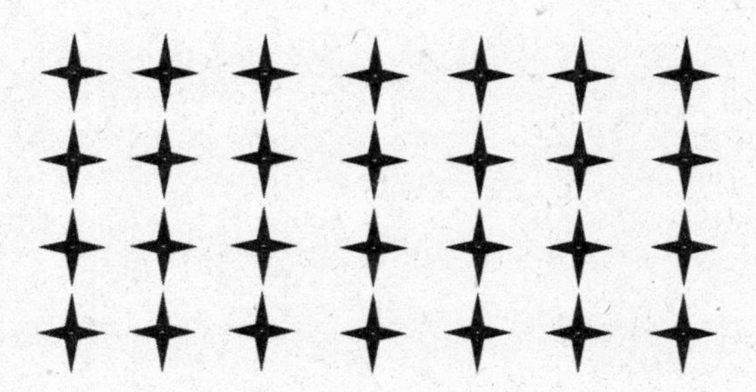

_______ = _______ × _______

_______ = _______ × _______

2. Karen says, "If I know $3 \times 8 = 24$, then I know the answer to 8×3." Explain why this is true.

Jocelyn says 7 fives has the same answer as 3 sevens + 2 sevens. Is she correct? Explain why or why not.

Read **Draw** **Write**

Name ______________________________ Date ______________

1. Each 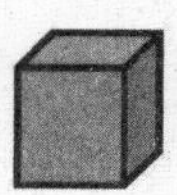 has a value of 7.

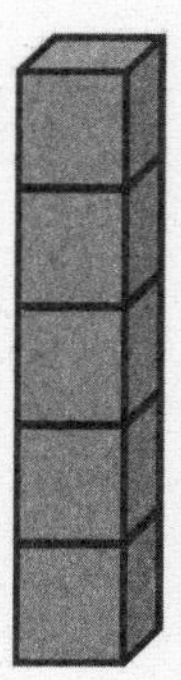

Unit form: 5 ____________

Facts: 5 × ______ = ______ × 5

Total = ______

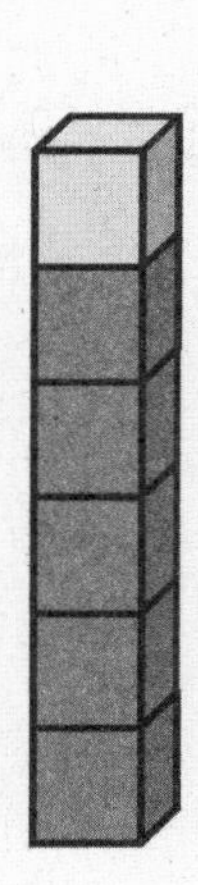

Unit form: 6 sevens = ______ sevens + ______ seven

= 35 + ______

= ______

Facts: ______ × ______ = ______

______ × ______ = ______

2. a. Each dot has a value of 8.

Unit form: 5 ____________

Facts: 5 × ______ = ______ × 5

Total = ______

b. Use the fact above to find 8 × 6. Show your work using pictures, numbers, or words.

3. An author writes 9 pages of her book each week. How many pages does she write in 7 weeks? Use a fives fact to solve.

---

4. Mrs. Gonzalez buys a total of 32 crayons for her classroom. Each pack contains 8 crayons. How many packs of crayons does Mrs. Gonzalez buy?

---

5. Hannah has $500. She buys a camera for $435 and 4 other items for $9 each. Now Hannah wants to buy speakers for $50. Does she have enough money to buy the speakers? Explain.

Name ______________________________ Date ________________

Use a fives fact to help you solve 7×6. Show your work using pictures, numbers, or words.

Twenty-four people line up to use the canoes at the park. Three people are assigned to each canoe. How many canoes are used?

Read **Draw** **Write**

Name ______________________________ Date ______________

1. Each equation contains a box representing the unknown. Find the value of the unknowns.

$5 \times 4 = \square$ $\square$ = _____

$21 \div 3 = \square$ $\square$ = _____

$24 \div \square = 4$ $\square$ = _____

$21 = \square \times 7$ $\square$ = _____

$32 = \square \times 8$ $\square$ = _____

$\square \div 10 = 7$ $\square$ = _____

$8 = 80 \div \square$ $\square$ = _____

$4 = 36 \div \square$ $\square$ = _____

$24 \div \square = 12$ $\square$ = _____

$8 = \square \div 3$ $\square$ = _____

$35 = 7 \times \square$ $\square$ = _____

2. Lonna buys 3 t-shirts for $8 each.

   a. What is the total amount Lonna spends on 3 t-shirts? Use a box to represent the total amount of money Lonna spends, and then solve the problem.

   b. If Lonna hands the cashier 3 ten dollar bills, how much change will she receive? Use a box in an equation to represent the change, and then find the value of the unknown.

3. Miss Potts used a total of 28 cups of flour to bake some bread. She used 4 cups of flour for each loaf of bread. How many loaves of bread did she bake? Represent the problem using multiplication and division sentences and a box for the unknown. Then, solve the problem.

_____ × _____ = ______

_____ ÷ _____ = ______

---

4. At a table tennis tournament, two games went on for a total of 32 minutes. One game took 12 minutes longer than the other. How long did it take to complete each game? Use different shapes to represent the unknowns. Solve the problem.

Name ______________________________ Date ______________

Find the value of the unknown in Problems 1–4.

1. $\square = 5 \times 9$

   $\square =$ ______

2. $30 \div 6 = \square$

   $\square =$ ______

3. $8 \times \square = 24$

   $\square =$ ______

4. $\square \div 4 = 7$

   $\square =$ ______

5. Mr. Strand waters his rose bushes for a total of 15 minutes. He waters each rose bush for 3 minutes. How many rose bushes does Mr. Strand water? Represent the problem using multiplication and division sentences and a box for the unknown. Then, solve the problem.

   _____ × _____ = _____

   _____ ÷ _____ = _____

Marshall puts 6 pictures on each of the 6 pages in his photo album. How many pictures does he put in the photo album in all?

Read **Draw** **Write**

Name ______________________________ Date ______________

1. Skip-count by six to fill in the blanks. Match each number in the count-by with its multiplication fact.

Count-by	Multiplication fact
6	9 × 6
____	6 × 6
18	4 × 6
____	7 × 6
30	2 × 6
36	1 × 6
____	3 × 6
48	10 × 6
____	5 × 6
60	8 × 6

2. Count by six to fill in the blanks below.

6, _______, _______, _______

Complete the multiplication equation that represents the final number in your count-by.

6 × _______ = _______

Complete the division equation that represents your count-by.

_______ ÷ 6 = _______

3. Count by six to fill in the blanks below.

6, _____, _____, _____, _____, _____, _____

Complete the multiplication equation that represents the final number in your count-by.

6 × _______ = _______

Complete the division equation that represents your count-by.

_______ ÷ 6 = _______

4. Mrs. Byrne's class skip-counts by six for a group counting activity. When she points up, they count up by six, and when she points down, they count down by six. The arrows show when she changes direction.

a. Fill in the blanks below to show the group counting answers.

↑0, 6, _____, 18, _____, ↓ _____, 12↑_____, 24, 30, _____ ↓30, 24, _____, ↑24, _____, 36, _____, 48

b. Mrs. Byrne says the last number that the class counts is the product of 6 and another number. Write a multiplication sentence and a division sentence to show she's right.

6 × _______ = 48 48 ÷ 6 = _______

5. Julie counts by six to solve 6 × 7. She says the answer is 36. Is she right? Explain your answer.

Name ______________________________ Date ______________

1. Sylvia solves 6×9 by adding $48 + 6$. Show how Sylvia breaks apart and bonds her numbers to complete the ten. Then, solve.

2. Skip-count by six to solve the following:

   a. $8 \times 6 =$ _____

   b. $54 \div 6 =$ _____

Gracie draws 7 rows of stars. In each row, she draws 4 stars. How many stars does Gracie draw in all? Use a box to represent the unknown and solve.

Read **Draw** **Write**

Name ______________________________ Date ______________

1. Skip-count by seven to fill in the blanks in the fish bowls. Match each count-by to its multiplication expression. Then, use the multiplication expression to write the related division fact directly to the right.

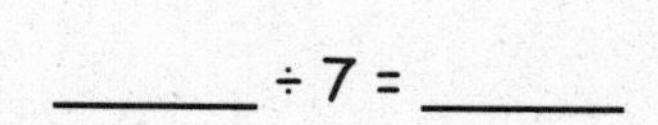
______ ÷ 7 = ______

______ ÷ 7 = ______

7 × 8
______ ÷ 7 = ______

______ ÷ 7 = ______

______ ÷ 7 = ______

______ ÷ 7 = ______

______ ÷ 7 = ______

______ ÷ 7 = ______

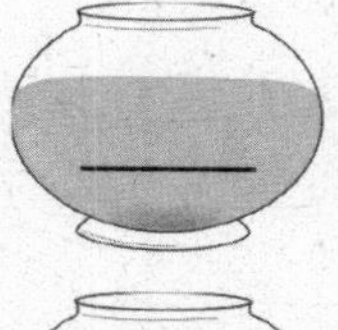

______ ÷ 7 = ______

______ ÷ 7 = ______

2. Complete the count-by seven sequence below. Then, write a multiplication equation and a division equation to represent each blank you filled in.

7, 14, _______, 28, _______, 42, _______, _______, 63, _______

a. _______ × 7 = _______ _______ ÷ 7 = _______

b. _______ × 7 = _______ _______ ÷ 7 = _______

c. _______ × 7 = _______ _______ ÷ 7 = _______

d. _______ × 7 = _______ _______ ÷ 7 = _______

e. _______ × 7 = _______ _______ ÷ 7 = _______

3. Abe says 3 × 7 = 21 because 1 seven is 7, 2 sevens are 14, and 3 sevens are 14 + 6 + 1, which equals 21. Why did Abe add 6 and 1 to 14 when he is counting by seven?

4. Molly says she can count by seven 6 times to solve 7 × 6. James says he can count by six 7 times to solve this problem. Who is right? Explain your answer.

Name ______________________________ Date ______________

Complete the count-by seven sequence below. Then, write a multiplication equation and a division equation to represent each number in the sequence.

7, 14, _____, 28, _____, 42, _____, _____, 63, _____

a. _____ × 7 = _____ _____ ÷ 7 = _____

b. _____ × 7 = _____ _____ ÷ 7 = _____

c. _____ × 7 = _____ _____ ÷ 7 = _____

d. _____ × 7 = _____ _____ ÷ 7 = _____

e. _____ × 7 = _____ _____ ÷ 7 = _____

f. _____ × 7 = _____ _____ ÷ 7 = _____

g. _____ × 7 = _____ _____ ÷ 7 = _____

h. _____ × 7 = _____ _____ ÷ 7 = _____

i. _____ × 7 = _____ _____ ÷ 7 = _____

j. _____ × 7 = _____ _____ ÷ 7 = _____

Name ______________________________ Date ________________

1. Match the words to the correct equation.

a number times 6 equals 30

7 times a number equals 42

6 times 7 equals a number

63 divided by a number equals 9

36 divided by a number equals 6

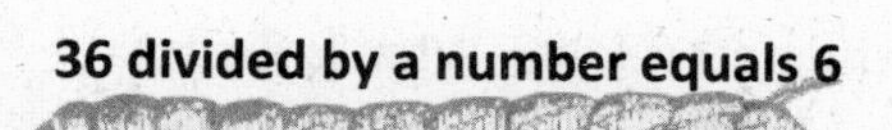

a number times 7 equals 21

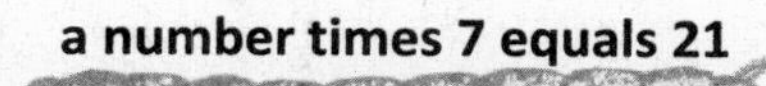

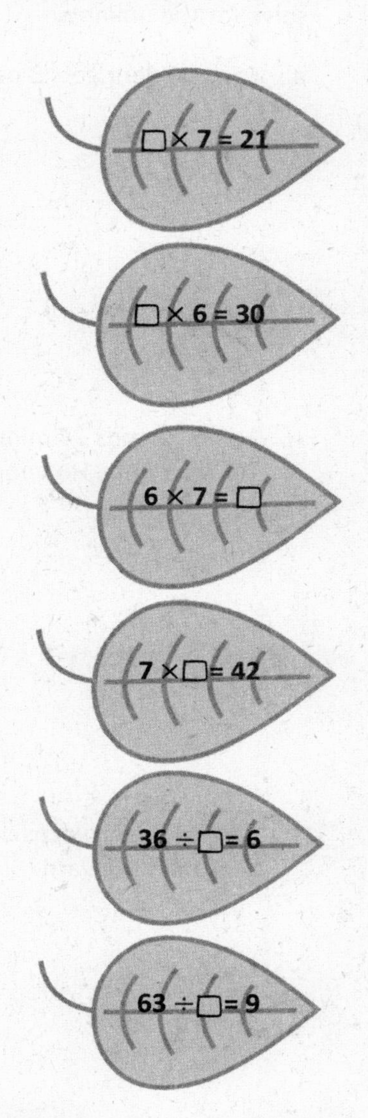

2. Write an equation with a box for the unknown to represent the strip diagram below, and solve for the unknown.

8	8	8	8	8	8

?

Equation: ______________________________

3. Model each problem with a drawing. Then, write an equation using a box to represent the unknown, and solve for the unknown.

   a. Each student gets 3 pencils. There are a total of 21 pencils. How many students are there?

   b. Henry spends 24 minutes practicing 6 different basketball drills. He spends the same amount of time on each drill. How much time does Henry spend on each drill?

   c. Jessica has 8 pieces of yarn for a project. Each piece of yarn is 6 centimeters long. What is the total length of the yarn?

   d. Ginny measures 6 milliliters of water into each beaker. She pours a total of 54 milliliters. How many beakers does Ginny use?

Name ______________________________ Date ______________

Model each problem with a drawing. Then, write an equation using a box to represent the unknown, and solve for the unknown.

1. Three boys and three girls each buy 7 bookmarks. How many bookmarks do they buy all together?

2. Seven friends equally share the cost of a $56 meal. How much does each person pay?

Richard has 2 cartons with 6 eggs in each. As he opens the cartons, he drops 2 eggs. How many unbroken eggs does Richard have left?

Read **Draw** **Write**

Name ______________________________ Date ______________

1. Solve.

a. $(12 - 4) + 6 =$ ______

b. $12 - (4 + 6) =$ ______

c. ______ $= 15 - (7 + 3)$

d. ______ $= (15 - 7) + 3$

e. ______ $= (3 + 2) \times 6$

f. ______ $= 3 + (2 \times 6)$

g. $4 \times (7 - 2) =$ ______

h. $(4 \times 7) - 2 =$ ______

i. ______ $= (12 \div 2) + 4$

j. ______ $= 12 \div (2 + 4)$

k. $9 + (15 \div 3) =$ ______

l. $(9 + 15) \div 3 =$ ______

m. $60 + (10 - 4) =$ ______

n. $(60 \div 10) - 4 =$ ______

o. ______ $= 35 + (10 \div 5)$

p. ______ $= (35 + 10) \div 5$

2. Use parentheses to make the equations true.

a. $16 - 4 + 7 = 19$	b. $16 - 4 + 7 = 5$
c. $2 = 22 - 15 + 5$	d. $12 = 22 - 15 + 5$
e. $3 + 7 \times 6 = 60$	f. $3 + 7 \times 6 = 45$
g. $5 = 10 \div 10 \times 5$	h. $50 = 100 \div 10 \times 5$
i. $26 - 5 \div 7 = 3$	j. $36 = 4 \times 25 - 16$

3. The teacher writes 24 ÷ 4 + 2 = ______ on the board. Chad says it equals 8. Samir says it equals 4. Explain how placing the parentheses in the equation can make both answers true.

4. Natasha solves the equation below by finding the sum of 5 and 12. Place the parentheses in the equation to show her thinking. Then, solve.

12 + 15 ÷ 3 = ________

5. Find two possible answers to the expression 7 + 3 × 2 by placing the parentheses in different places.

Name ______________________________ Date ______________

1. Use parentheses to make the equations true.

   a. $24 = 32 - 14 + 6$

   b. $12 = 32 - 14 + 6$

   c. $2 + 8 \times 7 = 70$

   d. $2 + 8 \times 7 = 58$

2. Marcos solves $24 \div 6 + 2 =$ _______. He says it equals 6. Iris says it equals 3. Show how the position of parentheses in the equation can make both answers true.

Name ______________________ Date ____________

Solve the following pairs of problems. Circle the pairs where both problems have the same answer.

1. a. $7 + (6 + 4)$

   b. $(7 + 6) + 4$

2. a. $(3 \times 2) \times 4$

   b. $3 \times (2 \times 4)$

3. a. $(2 \times 1) \times 5$

   b. $2 \times (1 \times 5)$

4. a. $(4 \times 2) \times 2$

   b. $4 \times (2 \times 2)$

5. a. $(3 + 2) \times 5$

   b. $3 + (2 \times 5)$

6. a. $(8 \div 2) \times 2$

   b. $8 \div (2 \times 2)$

7. a. $(9 - 5) + 3$

   b. $9 - (5 + 3)$

8. a. $(8 \times 5) - 4$

   b. $8 \times (5 - 4)$

Name ______________________________ Date ______________

1. Use the array to complete the equation.

a. $3 \times 12 =$ _____

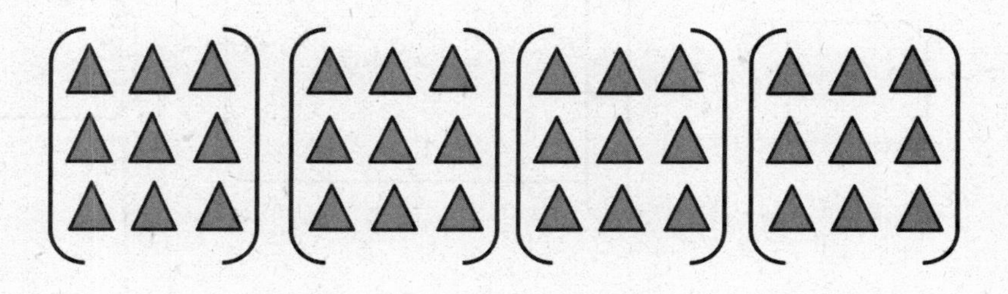

b. $(3 \times 3) \times 4$

$=$ _____ $\times 4$

$=$ _____

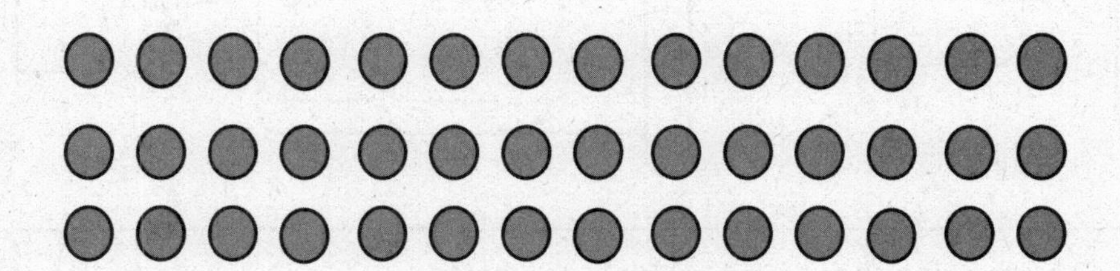

c. $3 \times 14 =$ ______

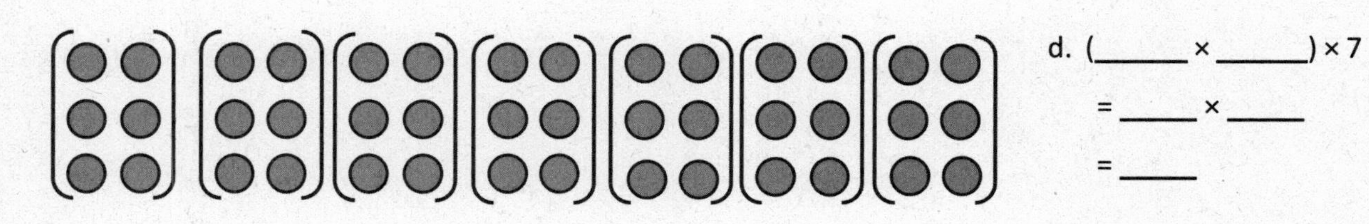

d. (______ × ______) × 7

$=$ _____ × _____

$=$ _____

2. Place parentheses in the equations to make it true. Then, solve. The first one has been done for you.

a.
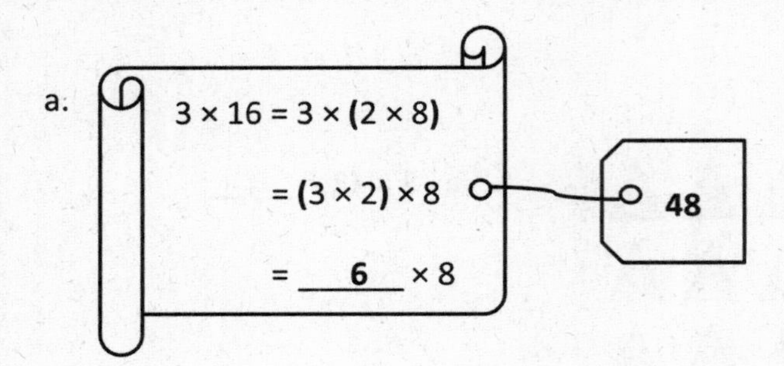
3 × 16 = 3 × (2 × 8)
= (3 × 2) × 8
= __6__ × 8

48

b.
2 × 14 = 2 × (2 × 7)
= (2 × 2) × 7
= ____ × 7

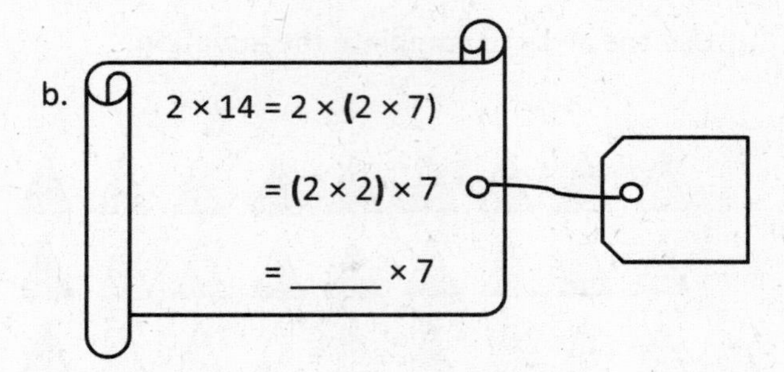

c.
3 × 12 = 3 × (3 × 4)
= 3 × 3 × 4
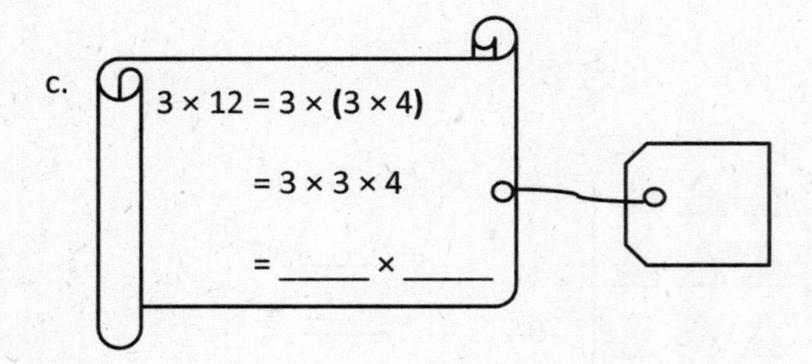

d.
3 × 14 = 3 × 2 × 7
= 3 × 2 × 7
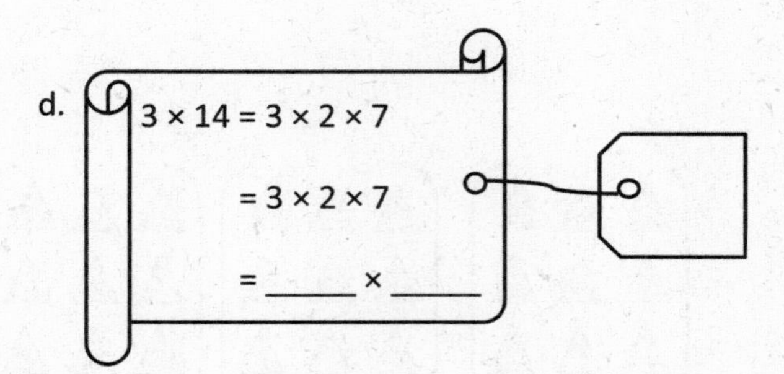

e.
15 × 3 = 5 × 3 × 3
= 5 × 3 × 3
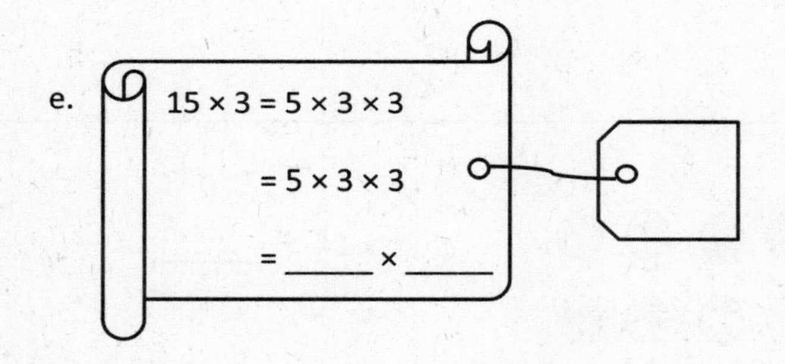

f.
15 × 2 = 5 × 3 × 2
= 5 × 3 × 2
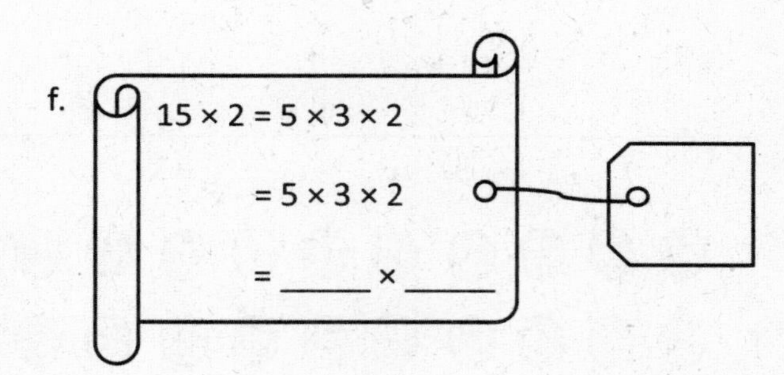

3. Charlotte finds the answer to 16 × 2 by thinking about 8 × 4. Explain her strategy.

Name ______________________________ Date ______________

Simplify to find the answer to 18 × 3. Show your work, and explain your strategy.

A scientist fills 5 test tubes with 9 milliliters of fresh water in each. She fills another 3 test tubes with 9 milliliters of salt water in each. How many milliliters of water does she use in all? Use the break apart and distribute strategy to solve.

Read **Draw** **Write**

Name ______________________________ Date ______________

1. Each  has a value of **9**. Find the value of each row. Then, add the rows to find the total.

a. **6 × 9 =** _____

5 × 9 = 45

1 × 9 = _____

6 × 9 = (5 + 1) × 9
= (5 × 9) + (1 × 9)
= 45 + _____
= _____

b. **7 × 9 =** _____

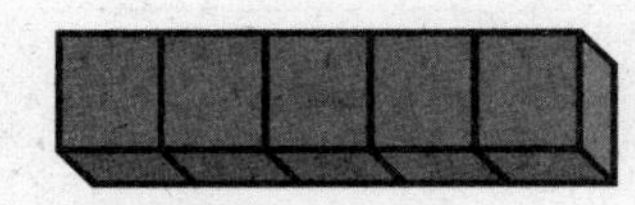

5 × 9 = 45

_____ × 9 = _____

7 × 9 = (5 + ____) × 9
= (5 × 9) + (____ × 9)
= 45 + _____
= _____

c. **8 × 9 =** _____

5 × 9 = _____

_____ × 9 = _____

8 × 9 = (5 + _____) × 9
= (5 × 9) + (_____ × _____)
= 45 + _____
= _____

d. **9 × 9 =** _____

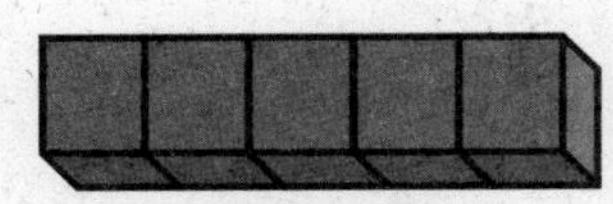

5 × 9 = _____

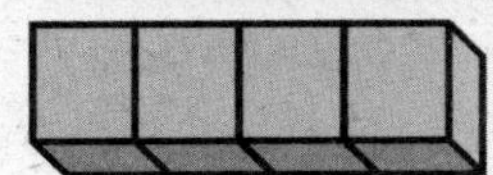

_____ × 9 = _____

9 × 9 = (5 + _____) × 9
= (5 × 9) + (_____ × _____)
= 45 + _____
= _____

2. Find the total value of the shaded blocks.

a. **9 × 6 =**

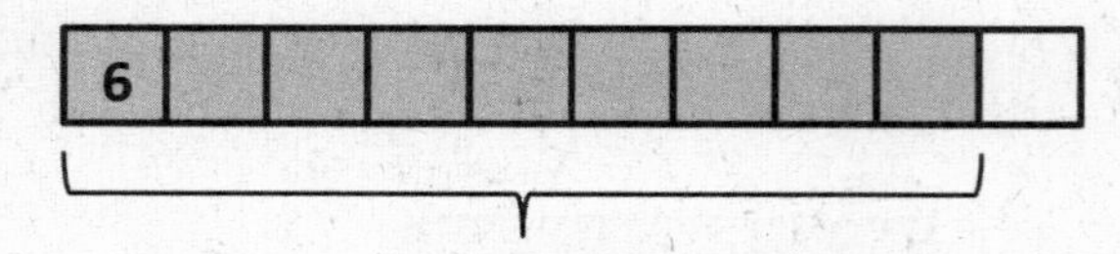

9 sixes = 10 sixes – 1 six

= _____ – 6

= _____

b. **9 × 7 =**

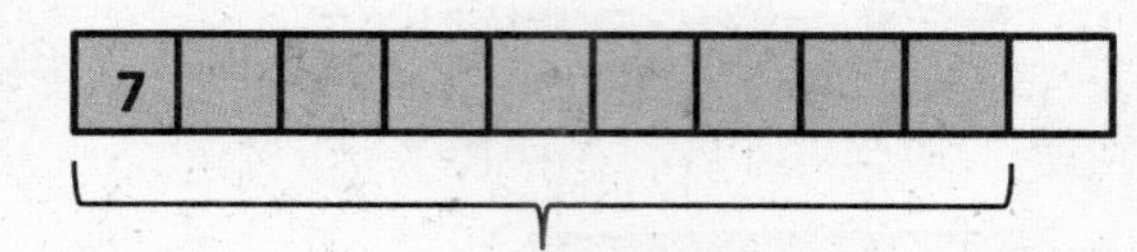

9 sevens = 10 sevens – 1 seven

= _____ – 7

= _____

c. **9 × 8 =**

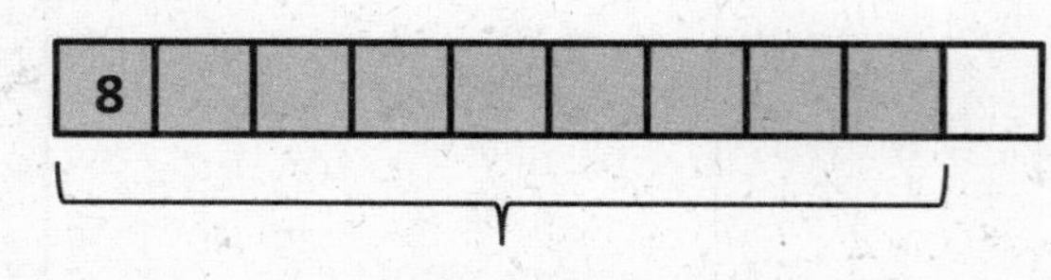

9 eights = 10 eights – 1 eight

= _____ – 8

= _____

d. **9 × 9 =**

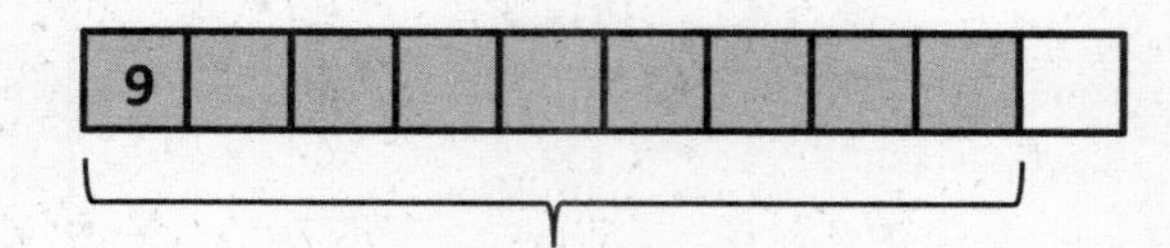

9 nines = 10 nines – 1 nine

= _____ – _____

= _____

3. Matt buys a pack of postage stamps. He counts 9 rows of 4 stamps. He thinks of 10 fours to find the total number of stamps. Show the strategy that Matt might have used to find the total number of stamps.

4. Match.

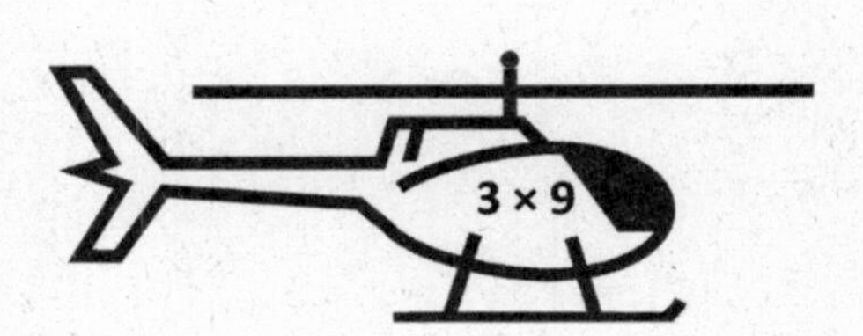

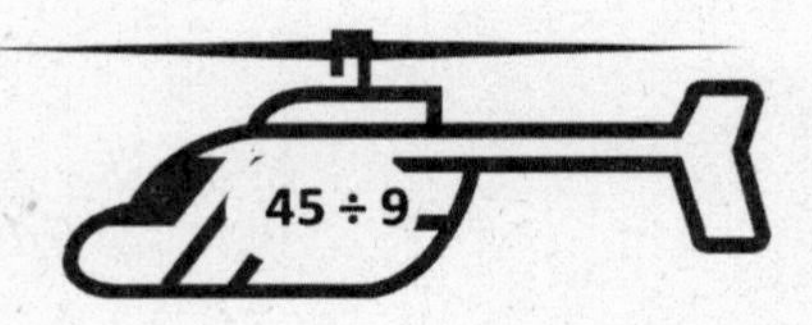

10

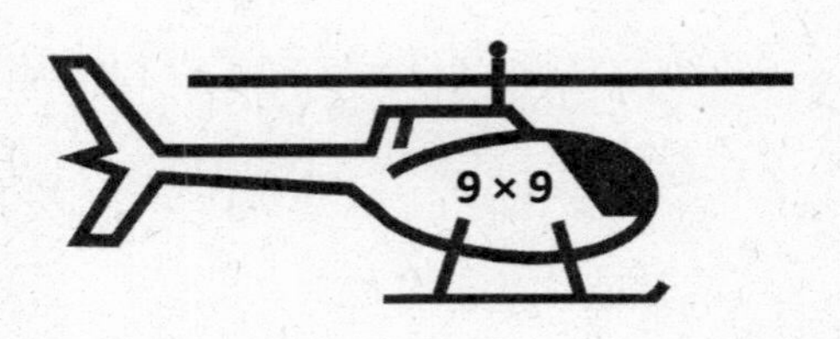

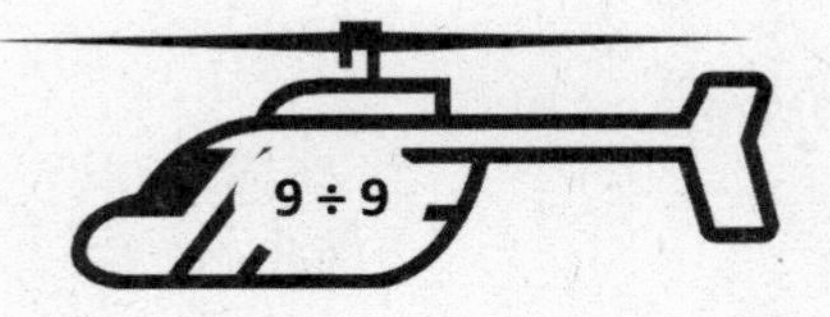

5

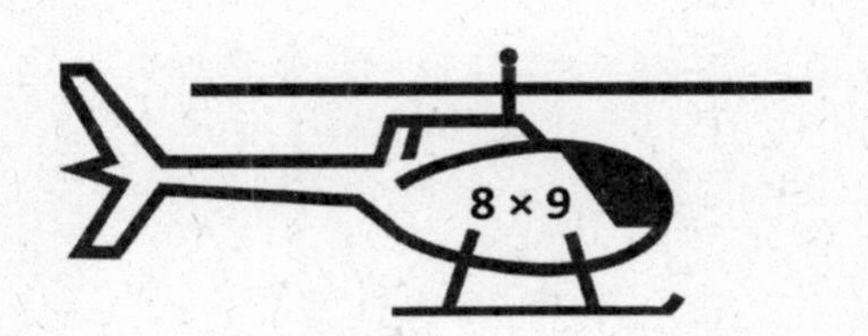

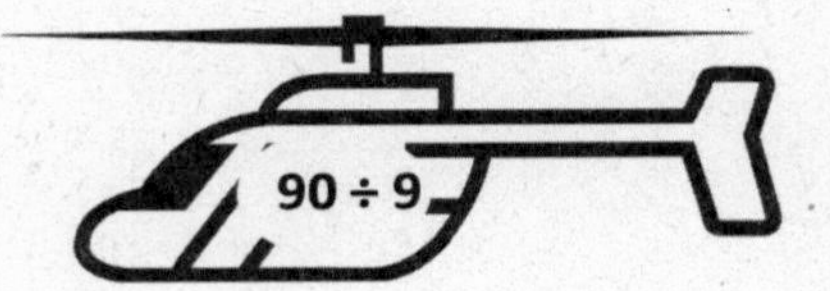

1

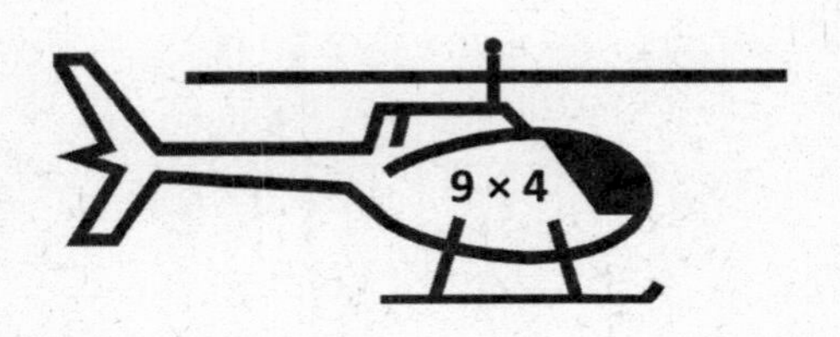

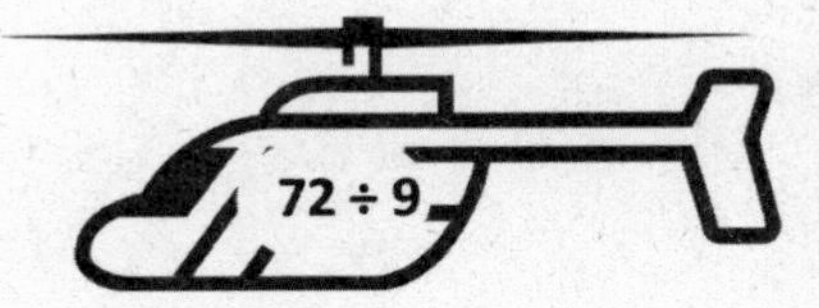

72

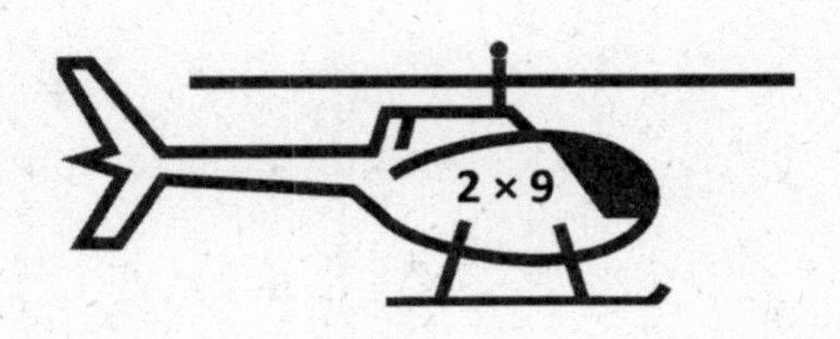

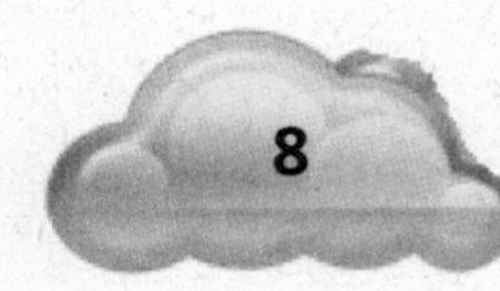

Name ______________________________ Date ______________

1. Each 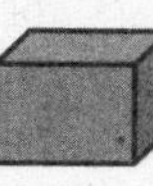 has a value of **9**. Complete the equations to find the total value of the tower of blocks.

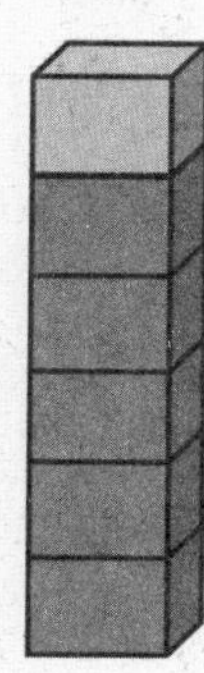

_____ × **9** = (5 + _____) × 9

= (5 × _____) + (_____ × _____)

= 45 + _____

= _____

2. Hector solves 9 × 8 by subtracting 1 eight from 10 eights. Draw a model, and explain Hector's strategy.

strip diagram

Name ______________________________ Date ______________

Write an equation, and use a question mark or a box to represent the unknown for Problems 1–6.

1. Mrs. Parson gave each of her grandchildren $9. She gave a total of $36. How many grandchildren does Mrs. Parson have?

2. Shiva pours 27 liters of water equally into 9 containers. How many liters of water are in each container?

3. Derek cuts 7 pieces of wire. Each piece is 9 meters long. What is the total length of the 7 pieces?

4. Aunt Deena and Uncle Chris share the cost of a pizza party with their 7 friends. The total cost of the pizzas is $63. If everyone shares the cost equally, how much does each person pay?

5. Cara bought 9 packs of beads. There are 10 beads in each pack. She always uses 30 beads to make each necklace. How many necklaces can she make if she uses all the beads?

6. There are 8 erasers in a set. Damon buys 9 sets. After giving some erasers away, Damon has 35 erasers left. How many erasers did he give away?

Name ______________________________ Date ______________

Use a question mark or a box to represent the unknown.

1. Mrs. Aquino pours 36 liters of water equally into 9 containers. How much water is in each container?

2. Marlon buys 9 packs of hot dogs. There are 6 hot dogs in each pack. After the barbeque, 35 hot dogs are left over. How many hot dogs were eaten?

Name ______________________________ Date ______________

1. Complete.

a. _____ × 1 = 6 b. _____ ÷ 7 = 0 c. 8 × _____ = 8 d. 9 ÷ _____ = 9

e. 0 ÷ 5 = _____ f. _____ × 0 = 0 g. 4 ÷ _____ = 1 h. _____ × 1 = 3

2. Match each equation with its solution.

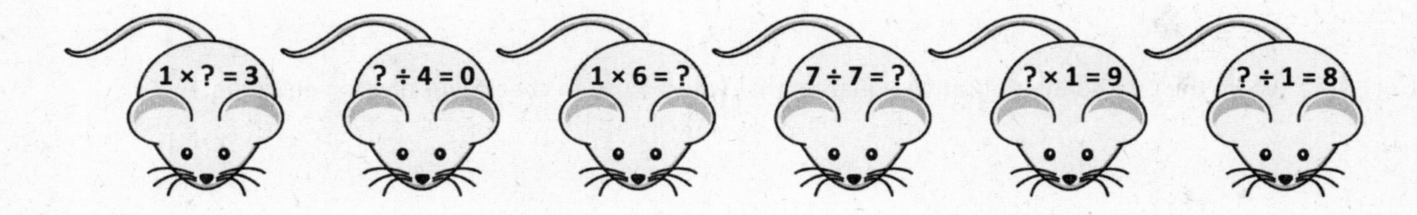

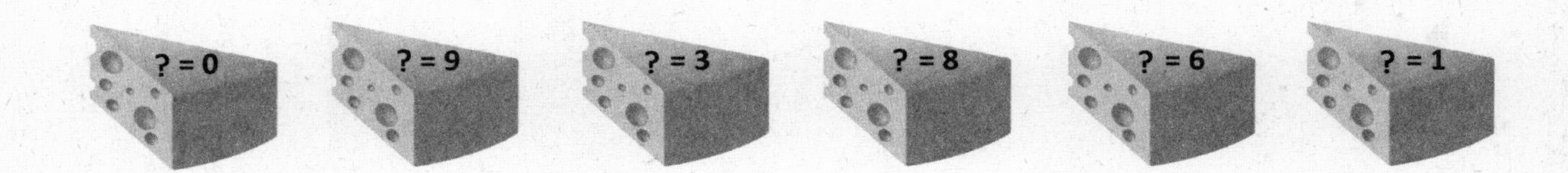

3. Let *n* be a number. Complete the blanks below with the products.

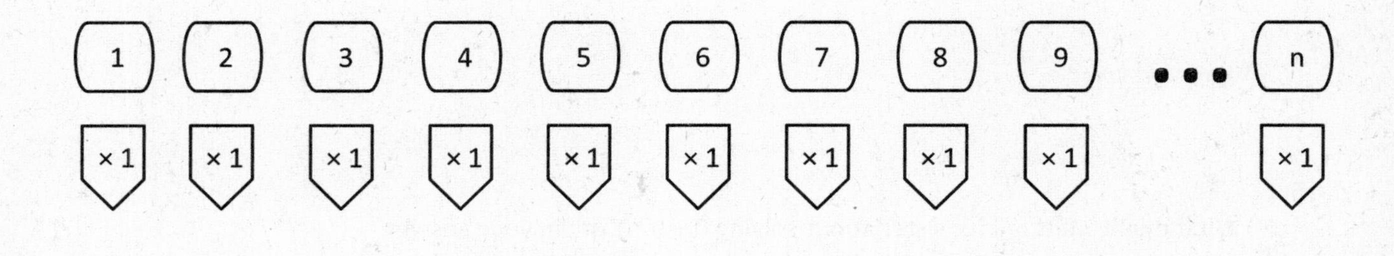

What pattern do you notice?

4. Josie says that any number divided by 1 equals that number.

   a. Write a division equation using n to represent Josie's statement.

   b. Use your equation from Part (a). Let $n = 6$. Write a new equation, and draw a picture to show that your equation is true.

   c. Write the related multiplication equation that you can use to check your division equation.

5. Matt explains what he learned about dividing with zero to his little sister.

   a. What might Matt tell his sister about solving $0 \div 9$? Explain your answer.

   b. What might Matt tell his sister about solving $8 \div 0$? Explain your answer.

   c. What might Matt tell his sister about solving $0 \div 0$? Explain your answer.

Name ______________________________ Date ______________

1. Complete.

a. _____ × 1 = 5

b. 6 × _____ = 6

c. _____ ÷ 7 = 0

d. 5 × _____ = 0

e. 1 = 9 ÷ _____

f. 8 = 1 × _____

2. Luis divides 8 by 0 and says it equals 0. Is he correct? Explain why or why not.

Henry's garden has 9 rows of squash plants. Each row has 8 squash plants. There is also 1 row with 8 watermelon plants. How many squash and watermelon plants does Henry have in all?

Read Draw Write

Name ______________________________ Date ______________

1. Write the products into the squares as fast as you can.

1 × 1	2 × 1	3 × 1	4 × 1	5 × 1	6 × 1	7 × 1	8 × 1
1 × 2	2 × 2	3 × 2	4 × 2	5 × 2	6 × 2	7 × 2	8 × 2
1 × 3	2 × 3	3 × 3	4 × 3	5 × 3	6 × 3	7 × 3	8 × 3
1 × 4	2 × 4	3 × 4	4 × 4	5 × 4	6 × 4	7 × 4	8 × 4
1 × 5	2 × 5	3 × 5	4 × 5	5 × 5	6 × 5	7 × 5	8 × 5
1 × 6	2 × 6	3 × 6	4 × 6	5 × 6	6 × 6	7 × 6	8 × 6
1 × 7	2 × 7	3 × 7	4 × 7	5 × 7	6 × 7	7 × 7	8 × 7
1 × 8	2 × 8	3 × 8	4 × 8	5 × 8	6 × 8	7 × 8	8 × 8

a. Color all the squares with even products orange. Can an even product ever have an odd factor?

b. Can an odd product ever have an even factor?

c. Everyone knows that $7 \times 4 = (5 \times 4) + (2 \times 4)$. Explain how this is shown in the table.

d. Use what you know to find the product of 7×16 or 8 sevens + 8 sevens.

2. In the table, only the products on the diagonal are shown.

   a. Label each product on the diagonal.

1 × 1					
	2 × 2				
		3 × 3			
			4 × 4		
				5 × 5	
					6 × 6

   b. Draw an array to match each expression in the table below. Then, label the number of squares you added to make each new array. The first two arrays have been done for you.

1 × 1	2 × 2	3 × 3	4 × 4	5 × 5	6 × 6

3 ____   ____   ____   ____   ____

c. What pattern do you notice in the number of squares that are added to each new array?

d. Use the pattern you discovered in Part (b) to prove this: 9 × 9 is the sum of the first 9 odd numbers.

Name ____________________ Date ____________

1. Use what you know to find the product of 8 × 12 or 6 eights + 6 eights.

2. Luis says 3 × 233 = 626. Use what you learned about odd times odd to explain why Luis is wrong.

Name ____ Date ____________________

Use the RDW process for each problem. Explain why your answer is reasonable.

1. Rose has 6 pieces of yarn that are each 9 centimeters long. Sasha gives Rose a piece of yarn. Now, Rose has a total of 81 centimeters of yarn. What is the length of the yarn that Sasha gives Rose?

2. Julio spends 29 minutes doing his spelling homework. He then completes each math problem in 4 minutes. There are 7 math problems. How many minutes does Julio spend on his homework in all?

3. Pearl buys 125 stickers. She gives 53 stickers to her little sister. Pearl then puts 9 stickers on each page of her album. If she uses all of her remaining stickers, on how many pages does Pearl put stickers?

4. Tanner's beaker had 45 milliliters of water in it at first. After each of his friends poured in 8 milliliters, the beaker contained 93 milliliters. How many friends poured water into Tanner's beaker?

5. Cora weighs 4 new, identical pencils and a ruler. The total weight of these items is 55 grams. She weighs the ruler by itself and it weighs 19 grams. How much does each pencil weigh?

Name ______________________________ Date ______________

Use the RDW process to solve. Explain why your answer is reasonable.

On Saturday, Warren swims laps in the pool for 45 minutes. On Sunday, he runs 8 miles. It takes him 9 minutes to run each mile. How long does Warren spend exercising over the weekend?

Mia has 152 beads. She uses some to make bracelets. Now there are 80 beads. If she uses 8 beads for each bracelet, how many bracelets does she make?

Read **Draw** **Write**

Name ______________________________ Date ______________

1. Use the disks to fill in the blanks in the equations.

a.

4×3 ones = _______ ones

4×3 = _______

b.

4×3 tens = _______ tens

4×30 = _______

2. Use the chart to complete the blanks in the equations.

tens	ones
	● ● ● ● ● ● ● ●

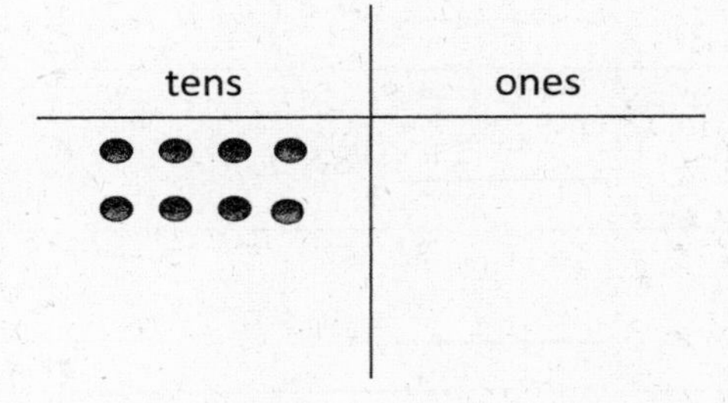

a. 2×4 ones = _______ ones

2×4 = _______

b. 2×4 tens = _______ tens

2×40 = _______

tens	ones
	● ● ● ● ● ● ● ● ● ● ● ● ● ● ●

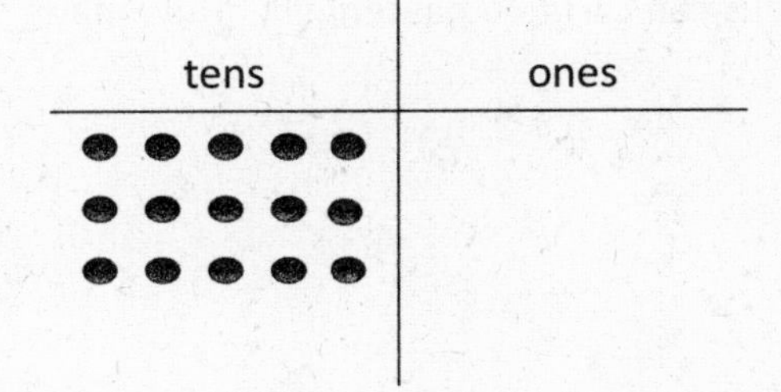

c. 3×5 ones = _______ ones

3×5 = _______

d. 3×5 tens = _______ tens

3×50 = _______

tens	ones

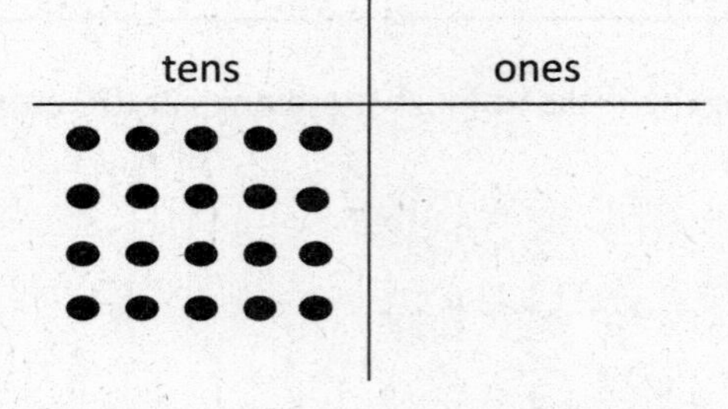

e. 4 × 5 ones = ________ ones

4 × 5 = ________

f. 4 × 5 ones = ________ ones

4 × 5 = ________

3. Fill in the blank to make the equation true.

a. __________ = 7 × 2	b. __________ tens = 7 tens × 2
c. __________ = 8 × 3	d. __________ tens = 8 tens × 3
e. __________ = 60 × 5	f. __________ = 4 × 80
g. 7 × 40 = __________	h. 50 × 8 = __________

4. A bus can carry 40 passengers. How many passengers can 6 buses carry? Model with a strip diagram.

Name ______________________________ Date ______________

1. Use the chart to complete the blanks in the equations.

tens	ones
	●●●●● ●●●●● ●●●●● ●●●●● ●●●●● ●●●●●

tens	ones
●●●●● ●●●●● ●●●●● ●●●●● ●●●●● ●●●●●	

6 × 5 ones = ______ ones

6 × 5 = ______

6 × 5 tens = ______ ones

6 × 50 = ______

2. A small plane has 20 rows of seats. Each row has 4 seats.

   a. Find the total number of seats on the plane.

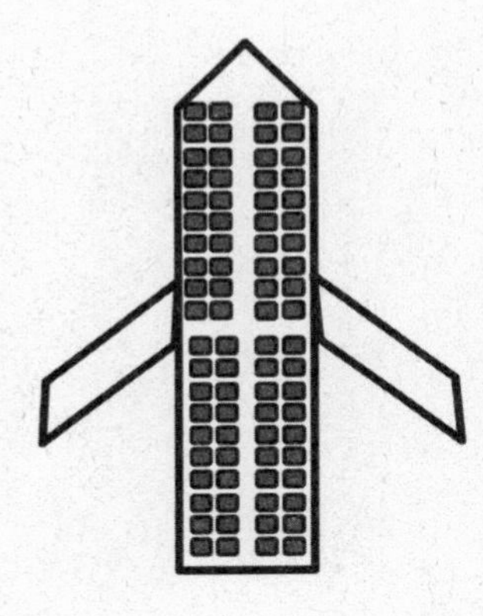

   b. How many seats are on 3 small planes?

Model 3 × 4 on a place value chart. Then, explain how the array can help you solve 30 × 4.

Read **Draw** **Write**

Name ______________________________ Date ______________

1. Use the chart to complete the equations. Then, solve. The first one has been done for you.

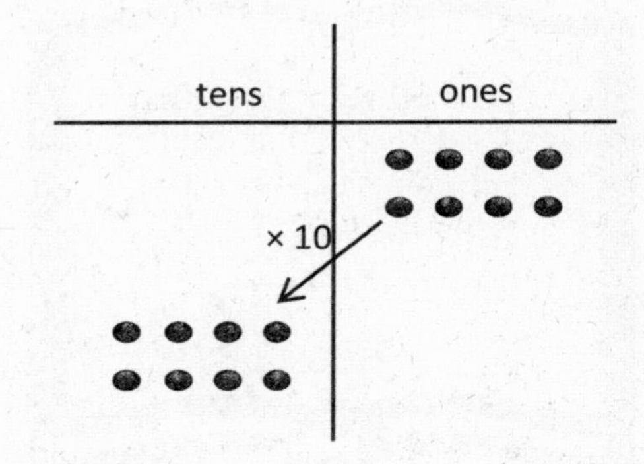

a. (2 × 4) × 10

= (8 ones) × 10

= 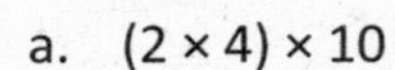 80

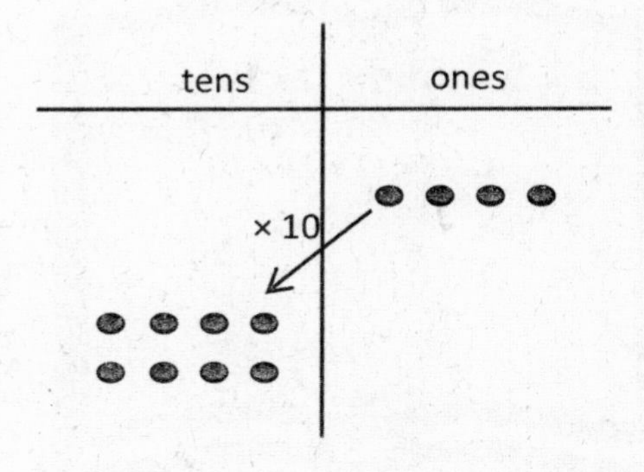

b. 2 × (4 × 10)

= 2 × (4 tens)

= ______

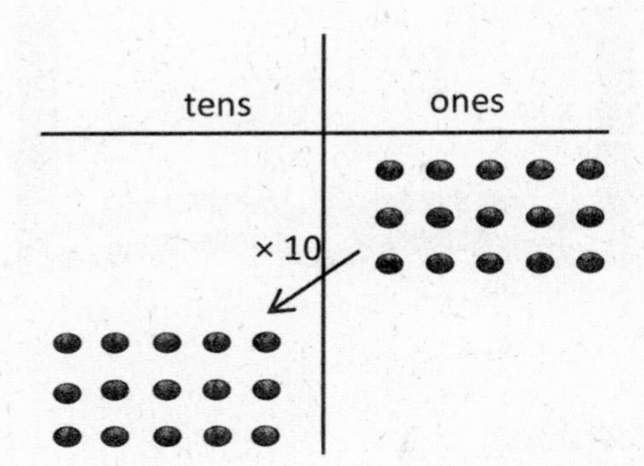

c. (3 × 5) × 10

= (_____ ones) × 10

= ______

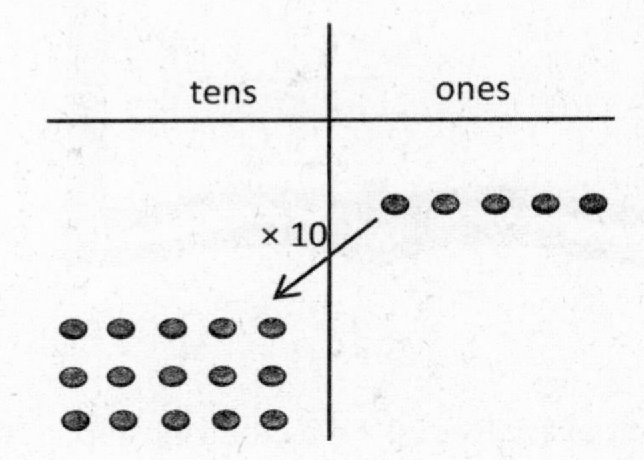

d. 3 × (5 × 10)

= 3 × (______ tens)

= ______

2. Place parentheses in the equations to find the related fact. Then, solve. The first one has been done for you.

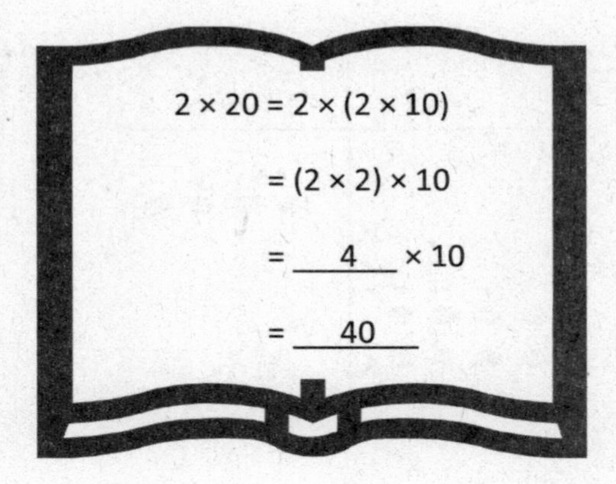

2 × 20 = 2 × (2 × 10)
= (2 × 2) × 10
= 4 × 10
= 40

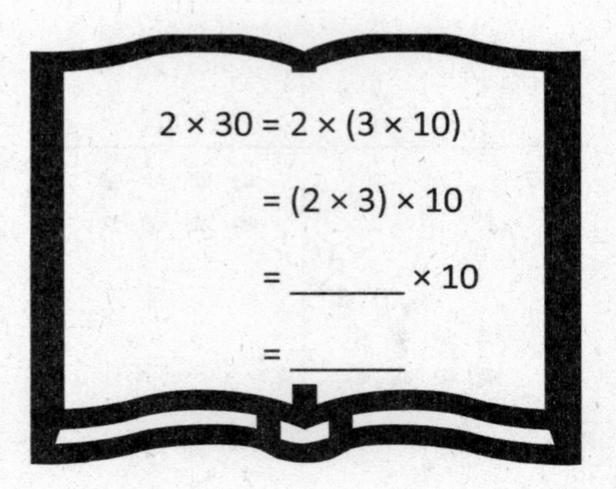

2 × 30 = 2 × (3 × 10)
= (2 × 3) × 10
= ______ × 10
= ______

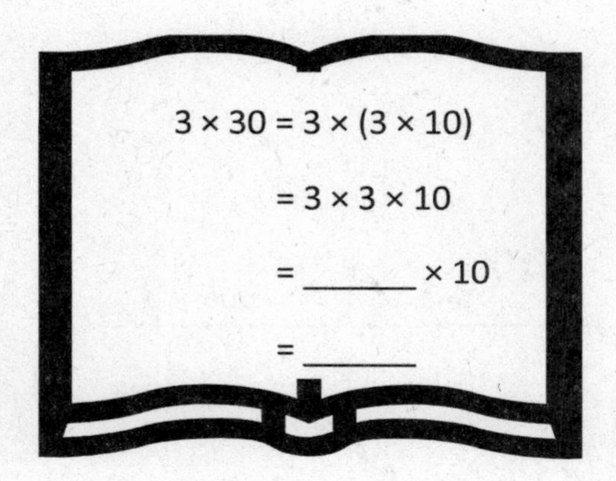

3 × 30 = 3 × (3 × 10)
= 3 × 3 × 10
= ______ × 10
= ______

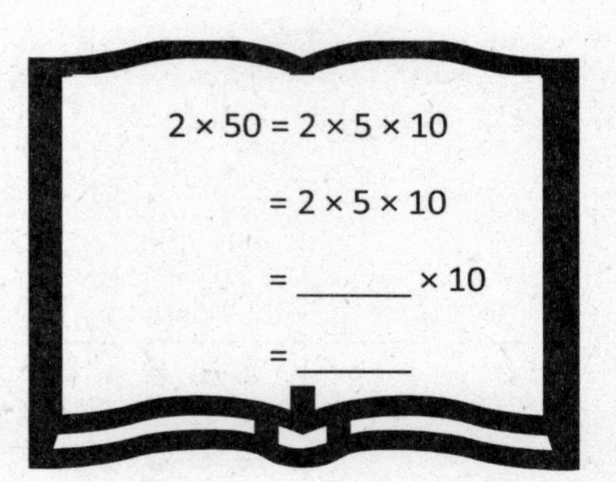

2 × 50 = 2 × 5 × 10
= 2 × 5 × 10
= ______ × 10
= ______

3. Gabriella solves 20 × 4 by thinking about 10 × 8. Explain her strategy.

Name ______________________________ Date ______________

1. Place parentheses in the equations to find the related fact. Then, solve.

a. $4 \times 20 = 4 \times 2 \times 10$

$= 4 \times 2 \times 10$

$=$ ____ $\times 10$

$=$ ____

b. $3 \times 30 = 3 \times 3 \times 10$

$= 3 \times 3 \times 10$

$=$ _____ $\times 10$

$=$

2. Jamila solves 20×5 by thinking about 10 tens. Explain her strategy.

Shondra has 3 boxes of books, each containing 4 books. Dan has 3 boxes of books, each containing 40 books. How many books do Shondra and Dan have in all?

Read **Draw** **Write**

Name ______________________________ Date ______________

1. Use your place value disks and charts to represent the following expressions. Record your work on the place value chart shown. Then write a matching expression, and record the partial products vertically. Problem (a) below is done for you.

a. 3 × 31

	10 10 10 10 10 10 10 10 10	1 1 1

$$\begin{array}{rl} 31 & \\ \times\ 3 & \\ \hline 3 & \leftarrow 3 \times 1 \text{ one} \\ +90 & \leftarrow 3 \times 3 \text{ tens} \\ \hline 93 & \leftarrow 3 \times 3 \text{ tens} + 3 \times 1 \text{ one} \end{array}$$

b. 3 × 23

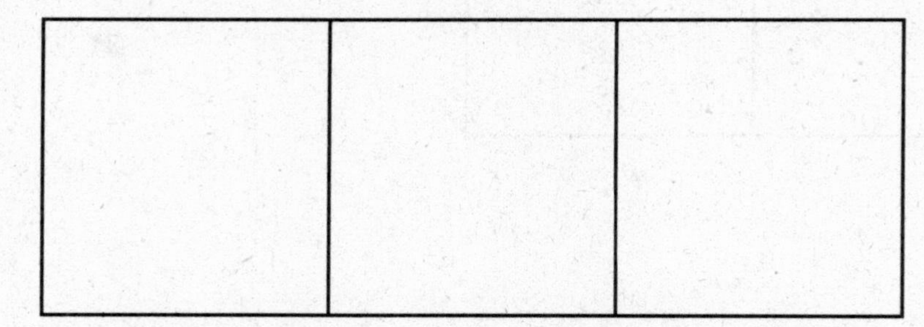

c. 4 × 23

d. 4 × 24

e. 3 × 25

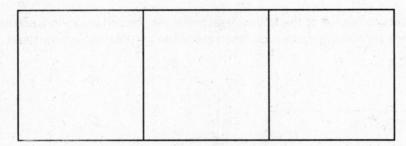

f. 3 × 42

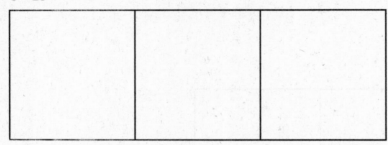

2. Tom says that knowing his multiplication facts helps him find products of larger numbers. He says that knowing that $3 \times 6 = 18$ and $3 \times 4 = 12$ helps him find the product of 3×64. What do you think Tom means? Explain your thinking in words, and justify your response by drawing place value disks on a chart and using partial products.

3. Julia and Luis are collecting pinecones for a craft project. On Saturday they collected 26 pinecones. On Sunday they collected 3 times as many pinecones as they collected on Saturday. How many pinecones did Julia and Luis collect altogether on Saturday and Sunday?

Name ______________________________ Date ________________

Use place value disks and a place value chart to solve these problems. Record the partial products vertically to the right of each expression.

1. 2 × 41

2. 2 × 35

hundreds place value chart

Keshon is collecting plastic water bottles as part of the third grade recycling project. His goal is to collect a total of 200 bottles. So far, he has 3 boxes of bottles, each containing 24 bottles. How many more plastic water bottles must Keshon collect to reach his goal?

Read **Draw** **Write**

Name ______________________________ Date ______________

1. Represent the expressions with disks. Write the problem vertically and record the partial products.

a. 2 × 23

hundreds	tens	ones
	• • • •	• • • • • •

```
   23
 ×  2
    6 ← 2 × 3 ones
 + 40 ← 2 × 2 tens
   46 ← 2 × 2 tens + 2 × 3 ones
```

b. 3 × 23

hundreds	tens	ones

c. 4 × 23

hundreds	tens	ones

d. 5 × 23

hundreds	tens	ones

2. Represent the expressions with disks. Write the problem vertically and record the partial products.

a. 3×34

hundreds	tens	ones

b. 3×53

hundreds	tens	ones

c. 6×24

hundreds	tens	ones

Name ______________________________ Date ______________

1. Represent the expressions with disks. Write the problem vertically and record the partial products.

   a. 5 × 14

hundreds	tens	ones

   b. 4 × 51

hundreds	tens	ones

Calculate the total amount of liquid in three test tubes if each test tube contains 42 mL of liquid.

Read **Draw** **Write**

Name ____________________________________ Date ____________________

1. Solve using each method.

Partial Products	Standard Algorithm	Partial Products	Standard Algorithm
a. $\begin{array}{r} 34 \\ \times \quad 4 \\ \hline \end{array}$	$\begin{array}{r} 34 \\ \times \quad 4 \\ \hline \end{array}$	b. $\begin{array}{r} 54 \\ \times \quad 3 \\ \hline \end{array}$	$\begin{array}{r} 54 \\ \times \quad 3 \\ \hline \end{array}$

2. Solve. Use the standard algorithm.

a. $\begin{array}{r} 62 \\ \times \quad 4 \\ \hline \end{array}$	b. $\begin{array}{r} 35 \\ \times \quad 6 \\ \hline \end{array}$	c. $\begin{array}{r} 64 \\ \times \quad 6 \\ \hline \end{array}$
d. $\begin{array}{r} 85 \\ \times \quad 4 \\ \hline \end{array}$	e. $\begin{array}{r} 36 \\ \times \quad 5 \\ \hline \end{array}$	f. $\begin{array}{r} 92 \\ \times \quad 6 \\ \hline \end{array}$

3. The product of 7 and 86 is ________.

4. 9 times as many as 47 is ________.

5. Kendra wants to make a lid for a small wooden box. She needs 5 pieces of wood each 14 centimeters long.

   How many centimeters of wood does she need?

6. One game costs $38. How much will 4 games cost?

Name ____________________ Date ____________

1. Solve using the standard algorithm.

a.	b.
$\begin{array}{r} 23 \\ \times \quad 9 \\ \hline \end{array}$	$\begin{array}{r} 74 \\ \times \quad 7 \\ \hline \end{array}$

2. Gerrie is 24 years old. Her grandmother is 3 times as old. How old is her grandmother?

Name ______________________________ Date ______________

Use the RDW process to solve each problem.

1. There are 60 seconds in 1 minute. Use a strip diagram to find the total number of seconds in 5 minutes and 45 seconds.

2. Lupe saves $32 each month for 4 months. Does she have enough money to buy the art supplies below? Explain why or why not.

3. Brad receives 5 cents for each can or bottle he recycles. How many cents does Brad earn if he recycles 50 cans and 32 bottles?

4. A box of 10 markers weighs 105 grams. If the empty box weighs 15 grams, how much does each marker weigh?

5. Mr. Perez buys 3 sets of cards. Each set comes with 18 striped cards and 12 polka dot cards. He uses 49 cards. How many cards does he have left?

6. Ezra earns $9 an hour working at a book store. She works for 7 hours each day on Mondays and Wednesdays. How much does Ezra earn each week?

Name ______________________________ Date ______________

Use the RDW process to solve.

Frederick buys a can of 3 tennis balls. The empty can weighs 20 grams, and each tennis ball weighs 62 grams. What is the total weight of the can with 3 tennis balls?

Grade 3
Module 4

Mara uses 15 square-centimeter tiles to make a rectangle. Ashton uses 9 square-centimeter tiles to make a rectangle.

a. Draw what Mara and Ashton's rectangles might look like.

b. Whose rectangle has a bigger area? How do you know?

Read **Draw** **Write**

Name ______________________________________ Date ________________

1. Use a ruler to measure the side lengths of the rectangle in centimeters. Mark each centimeter with a point and connect the points to show the square units. Then, count the squares you drew to find the total area.

Total area: ___________________________

2. Use a ruler to measure the side lengths of the rectangle in inches. Mark each inch with a point and connect the points to show the square units. Then, count the squares you drew to find the total area.

Total area: ___________________________

3. Mariana uses square centimeter tiles to find the side lengths of the rectangle below. Label each side length. Then, count the tiles to find the total area.

Total area: ___________________________

4. Each ☐ is 1 square centimeter. Saffron says that the side length of the rectangle below is 4 centimeters. Kevin says the side length is 5 centimeters. Who is correct? Explain how you know.

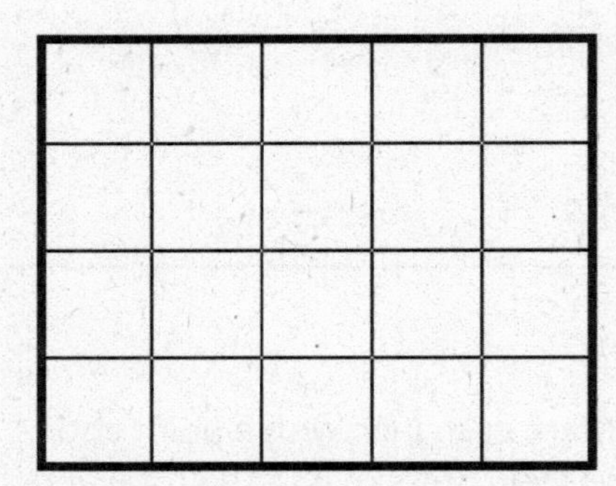

5. Use both square centimeter and square inch tiles to find the area of the rectangle below. Which works best? Explain why.

6. How does knowing side lengths A and B help you find side lengths C and D on the rectangle below?

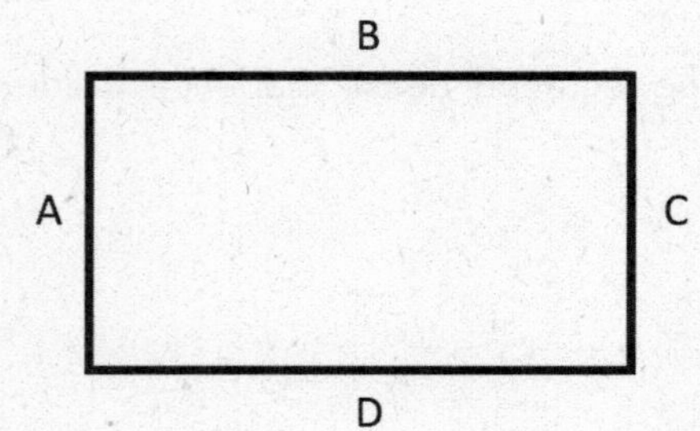

Name ______________________________ Date ______________

Label the side lengths of each rectangle. Then, match the rectangle to its total area.

a.

12 square centimeters

b.

5 square inches

c.

6 square centimeters

Candice uses square centimeter tiles to find the side lengths of a rectangle as shown on the right. She says the side lengths are 5 centimeters and 7 centimeters. Her partner, Luis, uses a ruler to check Candice's work and says that the side lengths are 5 centimeters and 6 centimeters. Who is right? How do you know?

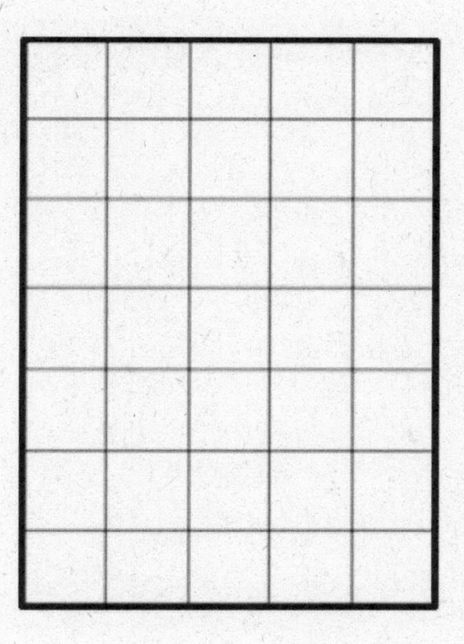

Read Draw Write

Name ____________________ Date ____________

1. Use the centimeter side of a ruler to draw in the tiles. Find the unknown side length or skip-count to find the unknown area. Then, complete the multiplication equations.

a. Area: **18** square centimeters.

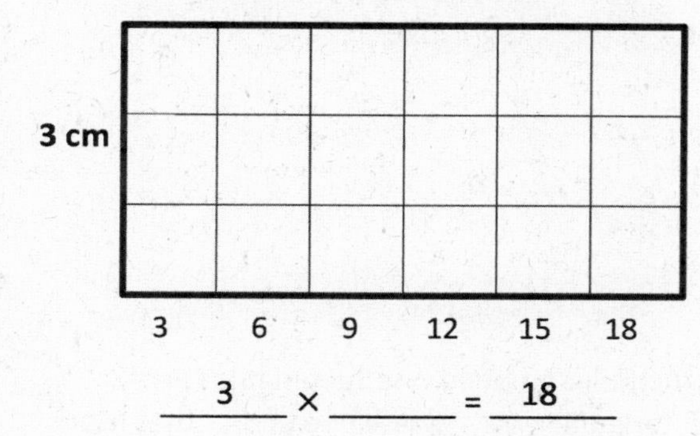

3 × ______ = 18

b. Area: **24** square centimeters.

______ × ______ = ______

c. Area: ______ square centimeters.

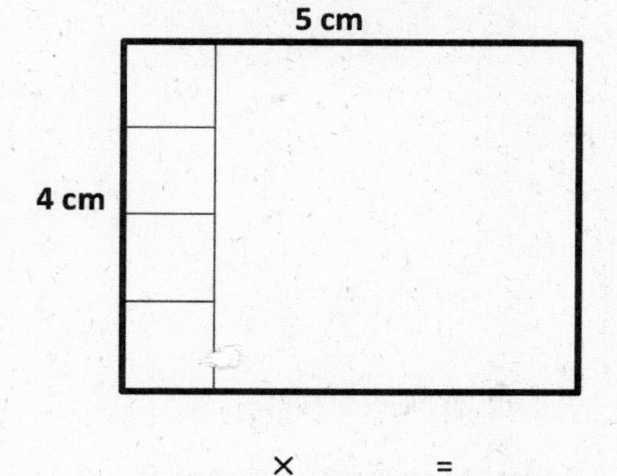

______ × ______ = ______

d. Area: **20** square centimeters.

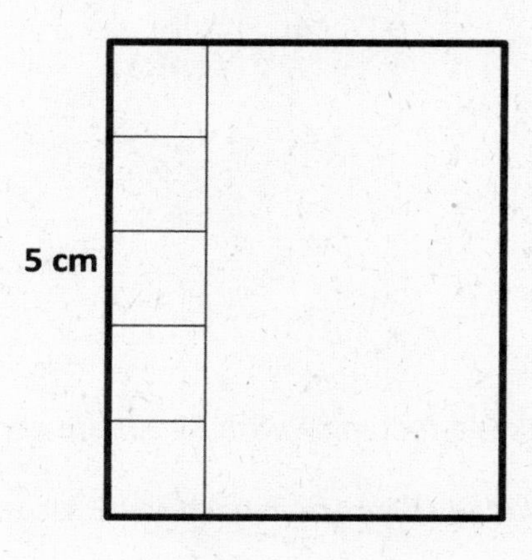

______ × ______ = ______

e. Area: **18** square centimeters.

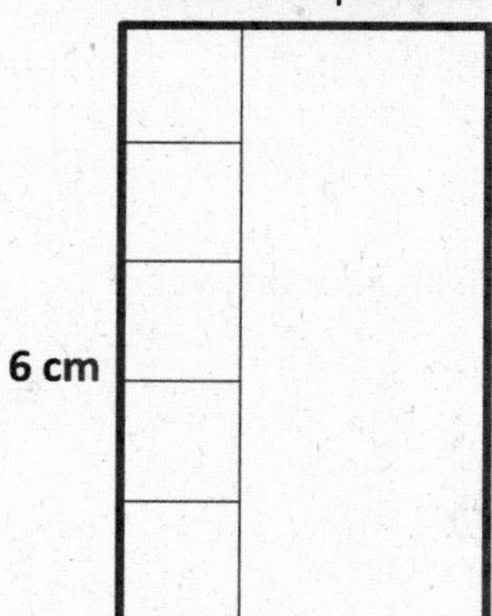

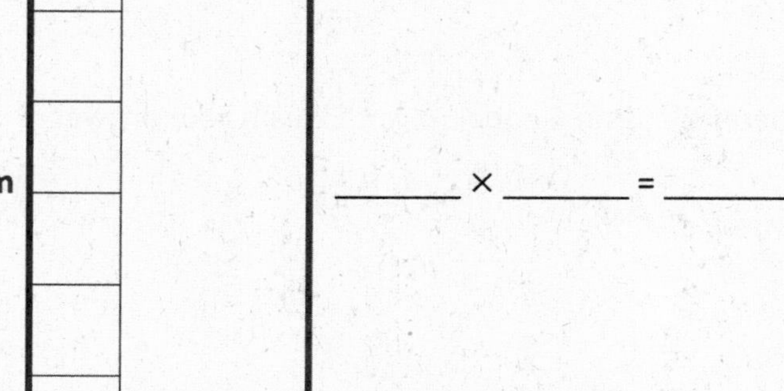

______ × ______ = ______

f. Area: ______ square centimeters.

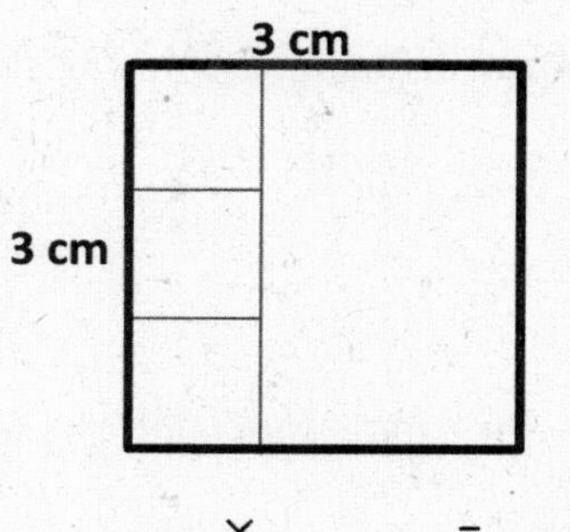

______ × ______ = ______

2. Lindsey makes a rectangle with 35 square inch tiles. She arranges the tiles in 5 equal rows. What are the side lengths of the rectangle? Use words, pictures, and numbers to support your answer.

3. Mark has a total of 24 square inch tiles. He uses 18 square inch tiles to build one rectangular array. He uses the remaining square inch tiles to build a second rectangular array. Draw two arrays that Mark might have made. Then, write multiplication sentences for each.

4. Leon makes a rectangle with 32 square centimeter tiles. There are 4 equal rows of tiles.

   a. How many tiles are in each row? Use words, pictures, and numbers to support your answer.

   b. Can Leon arrange all of his 32 square centimeter tiles into 6 equal rows? Explain your answer.

Name ______________________________ Date ______________

Darren has a total of 28 square centimeter tiles. He arranges them into 7 equal rows. Draw Darren's rectangle. Label the side lengths, and write a multiplication sentence to find the total area.

Huma has 4 bags of square inch tiles with 23 tiles in each bag. She uses them to measure the area of a rectangle on her homework. After covering the rectangle, Huma has 4 tiles left. What is the area of the rectangle?

Read **Draw** **Write**

Name ______________________________ Date ______________

1. Each ☐ represents 1 square centimeter. Draw to find the number of rows and columns in each array. Match it to its completed array. Then, fill in the blanks to make a true equation to find each array's area.

a. 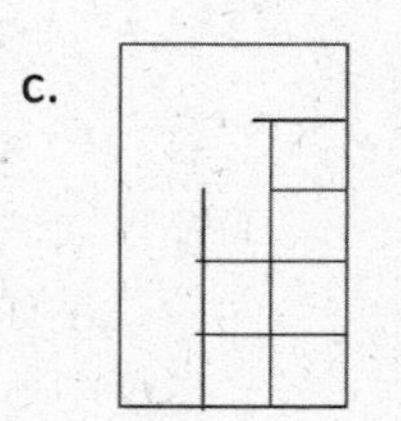 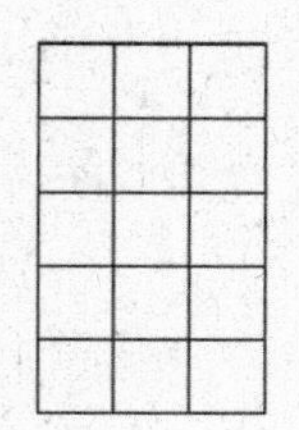 _____ cm × _____ cm = _____ sq cm

b. 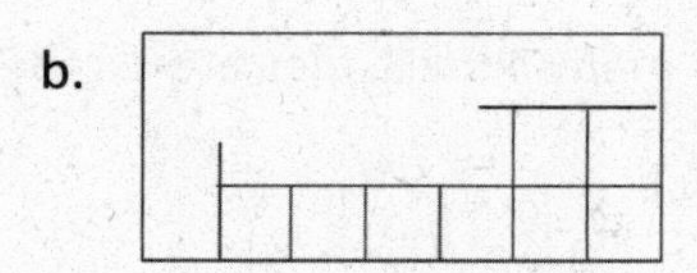 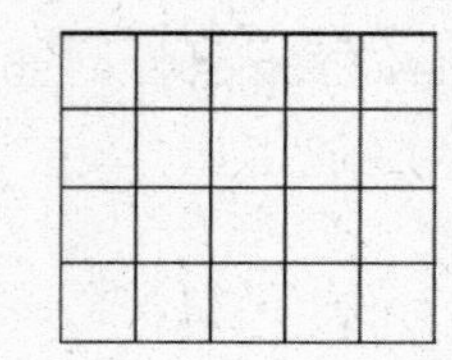 _____ cm × _____ cm = _____ sq cm

c. 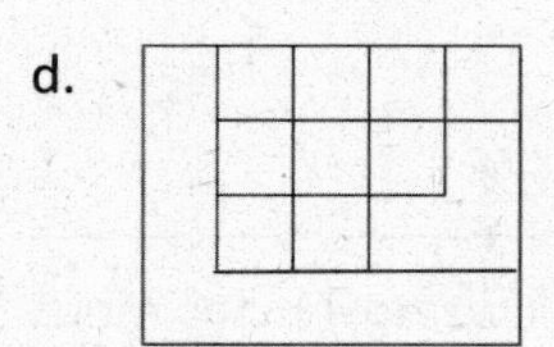 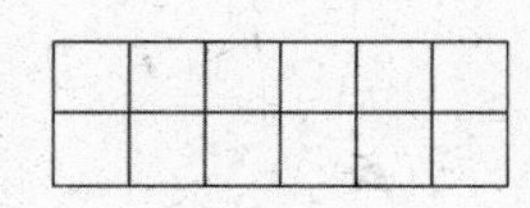 _____ cm × _____ cm = _____ sq cm

d. 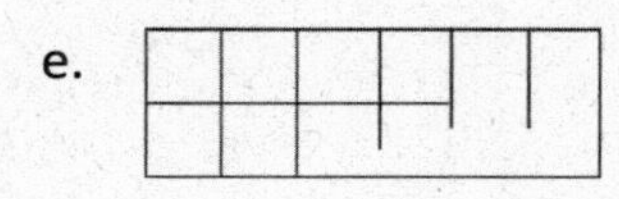 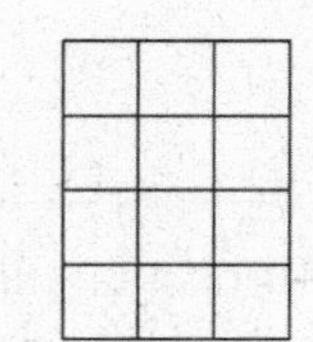 _____ cm × _____ cm = _____ sq cm

e. 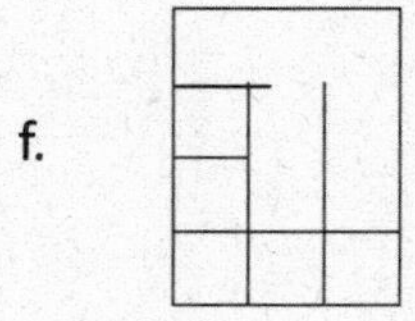  _____ cm × _____ cm = _____ sq cm

f. _____ cm × _____ cm = _____ sq cm

2. Sheena skip-counts by sixes to find the total square units in the rectangle below. She says there are 42 square units. Is she right? Explain your answer.

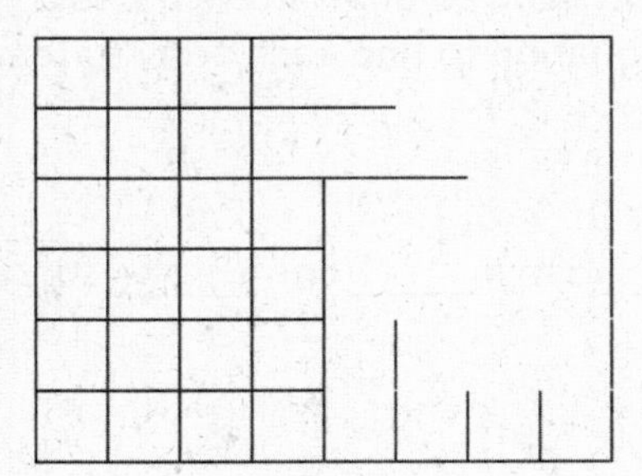

3. The tile floor in Brandon's living room has a rug on it as shown below. How many square tiles are on the floor, including the tiles under the rug?

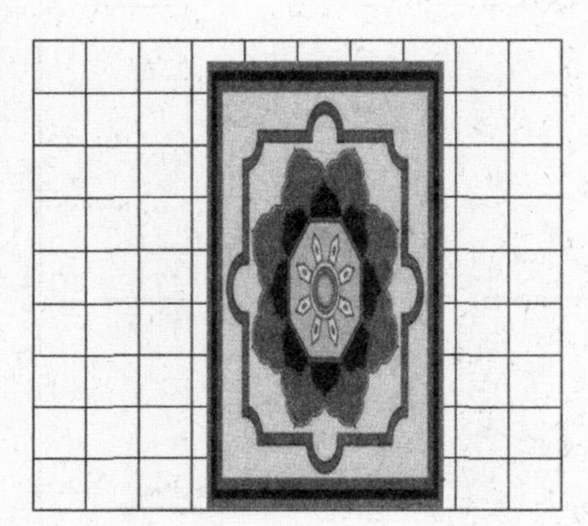

4. Abdul is creating a stained glass window with square inch glass tiles as shown below. How many more square inch glass tiles does Abdul need to finish his glass window? Explain your answer.

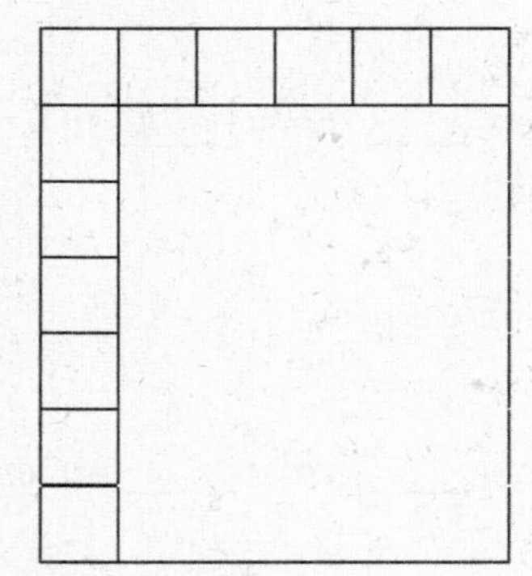

Name ______________________________ Date ________________

The tiled floor in Cayden's dining room has a rug on it as shown below. How many square tiles are on the floor, including the tiles under the rug?

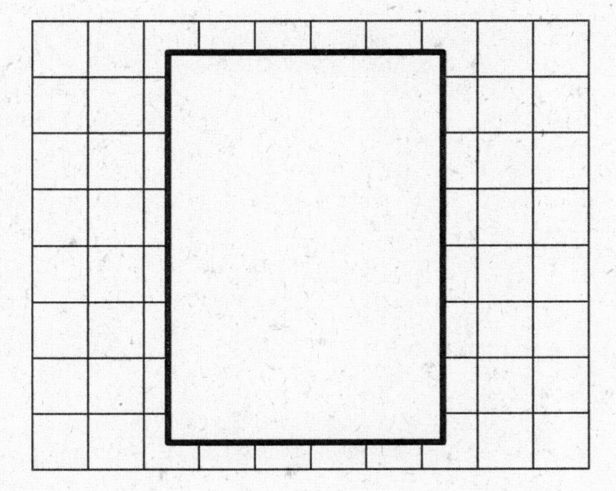

array 1

array 2

Lori wants to replace the square tiles on her wall. The square tiles are sold in boxes of 8 square tiles. Lori buys 6 boxes of tiles. Does she have enough to replace all of the tiles, including the tiles under the painting? Explain your answer.

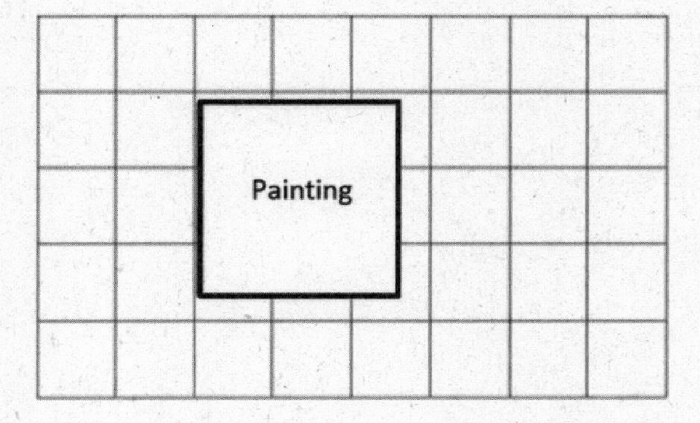

Read Draw Write

Name ________________________________ Date ________________

1. Use a straight edge to draw a grid of equal size squares within the rectangle. Find and label the side lengths. Then, multiply the side lengths to find the area.

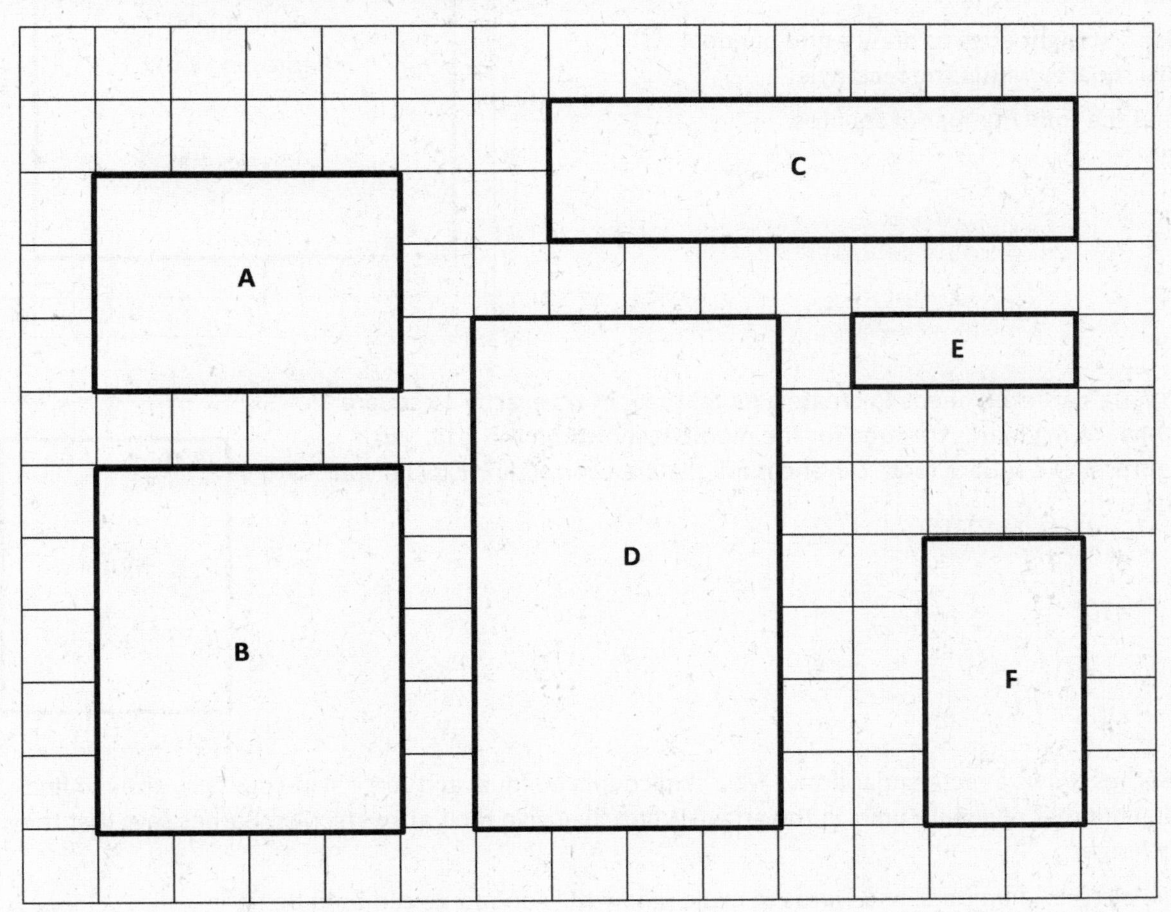

a. Area A:

____ units × ____ units = ____ square units

b. Area D:

____ units × ____ units = ____ square units

c. Area B:

____ units × ____ units = ____ square units

d. Area E:

____ units × ____ units = ____ square units

e. Area C:

____ units × ____ units = ____ square units

f. Area F:

____ units × ____ units = ____ square units

2. The area of Benjamin's bedroom floor is shown on the grid to the right. Each ☐ represents 1 square foot. How many total square feet is Benjamin's floor?

   a. Label the side lengths.

   b. Use a straight edge to draw a grid of equal size squares within the rectangle.

   c. Find the total number of squares.

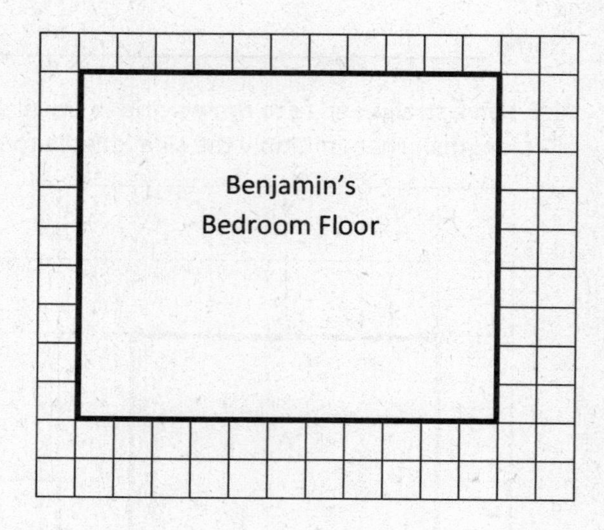

3. Mrs. Young's art class needs to create a mural that covers exactly 35 square feet. Mrs. Young marks the area for the mural as shown on the grid. Each ☐ represents 1 square foot. Did she mark the area correctly? Explain your answer.

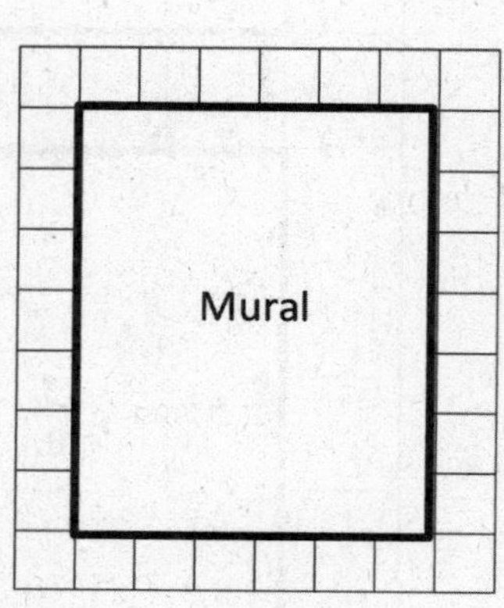

4. Mrs. Barnes draws a rectangular array. Mila skip-counts by fours and Jorge skip-counts by sixes to find the total number of square units in the array. When they give their answers, Mrs. Barnes says that they are both right.

   a. Use pictures, numbers, and words to explain how Mila and Jorge can both be right.

   b. How many square units might Mrs. Barnes' array have had?

Name ______________________________ Date ________________

1. Label the side lengths of Rectangle A on the grid below. Use a straight edge to draw a grid of equal size squares within Rectangle A. Find the total area of Rectangle A.

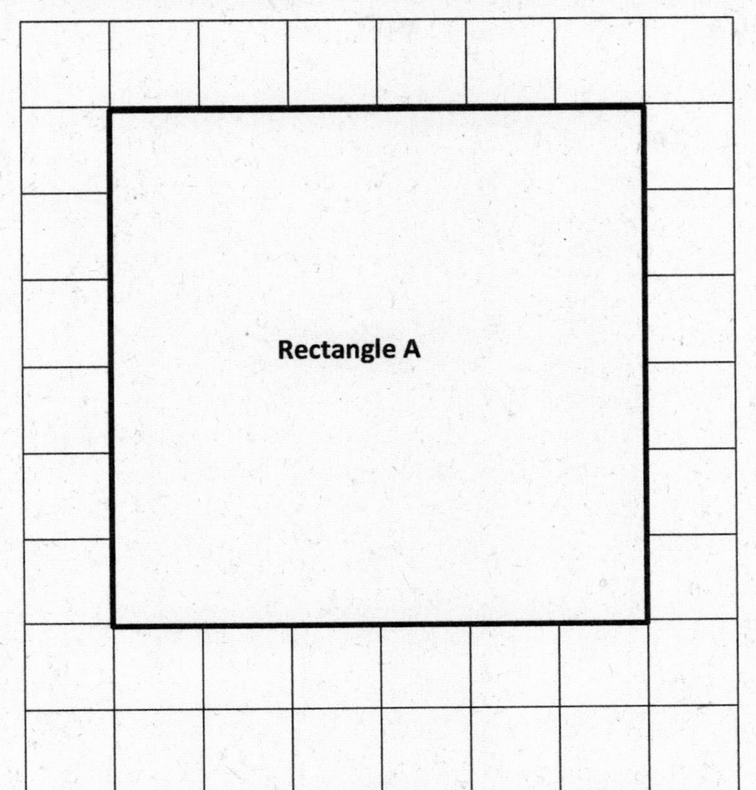

Area: ________ square units

2. Mark makes a rectangle with 36 square centimeter tiles. Gia makes a rectangle with 36 square inch tiles. Whose rectangle has a bigger area? Explain your answer.

area model

Marnie and Connor both skip-count square units to find the area of the same rectangle. Marnie counts, "3, 6, 9, 12, 15, 18, 21." Connor counts, "7, 14, 21." Draw what the rectangle might look like, and then label the side lengths and find the area.

Read **Draw** **Write**

Name ______________________________ Date ______________

1. Write a multiplication equation to find the area of each rectangle.

a. 7 ft by 4 ft — Area: ______ sq ft

_______ × _______ = _______

b. 7 ft by 8 ft — Area: ______ sq ft

_______ × _______ = _______

c. 6 ft by 6 ft — Area: ______ sq ft

_______ × _______ = _______

2. Write a multiplication equation and a division equation to find the unknown side length for each rectangle.

a. ______ ft by 9 ft — Area = 72 sq ft

_______ × _______ = _______

_______ ÷ _______ = _______

b. ______ ft by 3 ft — Area = 15 sq ft

_______ × _______ = _______

_______ ÷ _______ = _______

c. 4 ft by ______ ft — Area = 28 sq ft

_______ × _______ = _______

_______ ÷ _______ = _______

3. On the grid below, draw a rectangle that has an area of 42 square units. Label the side lengths.

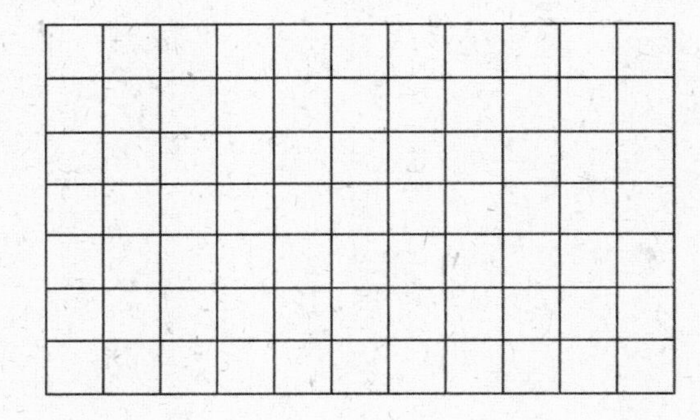

4. Ursa draws a rectangle that has side lengths of 9 centimeters and 6 centimeters. What is the area of the rectangle? Explain how you found your answer.

5. Eliza's bedroom measures 6 feet by 7 feet. Her brother's bedroom measures 5 feet by 8 feet. Eliza says their rooms have the same exact floor area. Is she right? Why or why not?

6. Cliff draws a rectangle with a side length of 6 inches and an area of 24 square inches. What is the other side length? How do you know?

Name ______________________________ Date ______________

1. Write a multiplication equation to find the area of the rectangle below.

9 inches

3 inches

Area: _____ sq in

_______ × _______ = _______

2. Write a multiplication equation and a division equation to find the unknown side length for the rectangle below.

_____ inches

6 inches

Area: 54 sq in

_______ × _______ = _______

_______ ÷ _______ = _______

grid

Mario plans to completely cover his 8-inch by 6-inch piece of cardboard with square inch tiles. He has 42 square inch tiles. How many more square inch tiles does Mario need to cover the cardboard without any gaps or overlap? Explain your answer.

Read **Draw** **Write**

Name ______________________________ Date ______________

1. Cut the grid into 2 equal rectangles.

   a. Draw and label the side lengths of the 2 rectangles.

   b. Write an equation to find the area of 1 of the rectangles.

   c. Write an equation to show the total area of the 2 rectangles.

2. Place your 2 equal rectangles side by side to create a new, longer rectangle.

   a. Draw an area model to show the new rectangle. Label the side lengths.

   b. Find the total area of the longer rectangle.

3. Furaha and Rahema use square tiles to make the rectangles shown below.

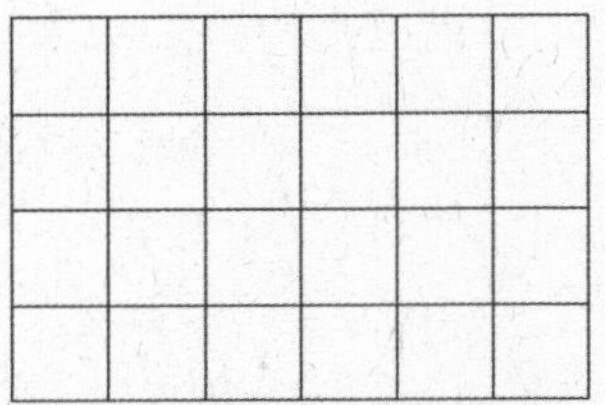
Furaha's Rectangle

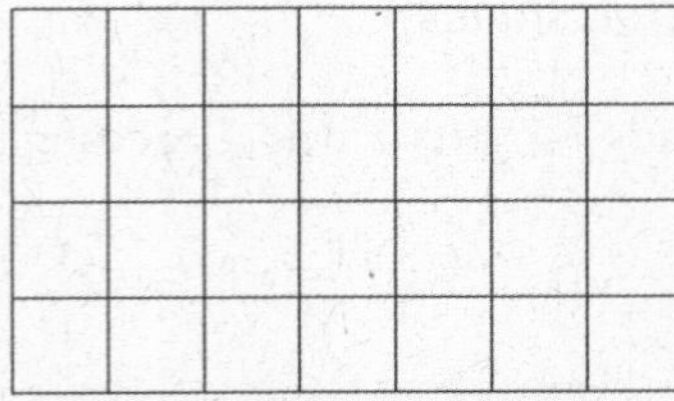
Rahema's Rectangle

a. Label the side lengths on the rectangles above, and find the area of each rectangle.

b. Furaha pushes his rectangle next to Rahema's rectangle to form a new, longer rectangle. Draw an area model to show the new rectangle. Label the side lengths.

c. Rahema says the area of the new, longer rectangle is 52 square units. Is she right? Explain your answer.

4. Kiera says she can find the area of the long rectangle below by adding the areas of Rectangles A and B. Is she right? Why or why not?

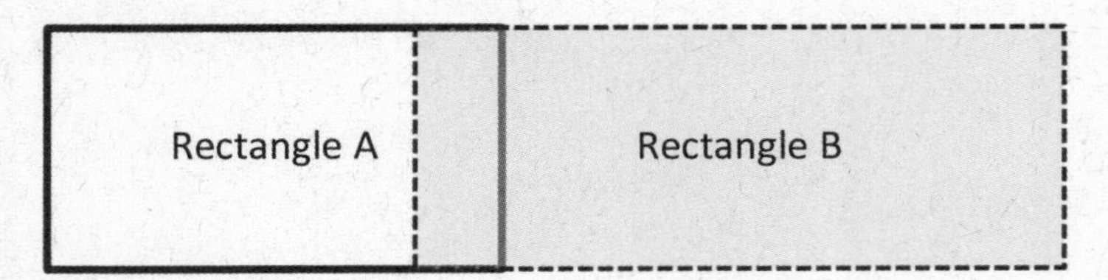

Name ______________________________ Date ______________

Lamar uses square tiles to make the 2 rectangles shown below.

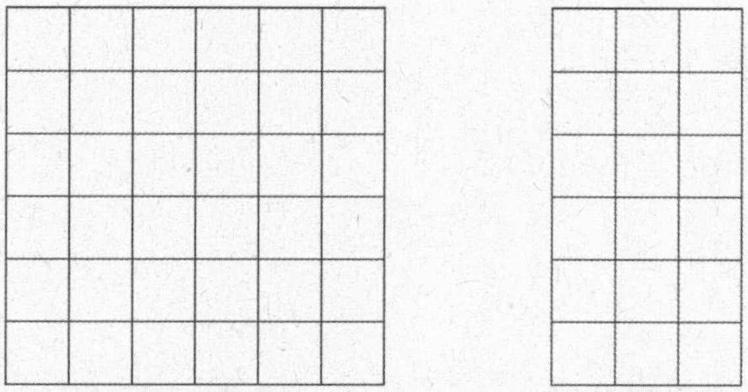

Rectangle A

Rectangle B

1. Label the side lengths of the 2 rectangles.

2. Write equations to find the areas of the rectangles.

   Area of Rectangle A: ______________ Area of Rectangle B: ______________

3. Lamar pushes Rectangle A next to Rectangle B to make a bigger rectangle. What is the area of the bigger rectangle? How do you know?

small centimeter grid

Sonya folds a 6-inch by 6-inch piece of paper into 4 equal parts (shown below). What is the area of 1 of the parts? Write a sentence using "times as much" that relates the area of the whole paper to the area of one part.

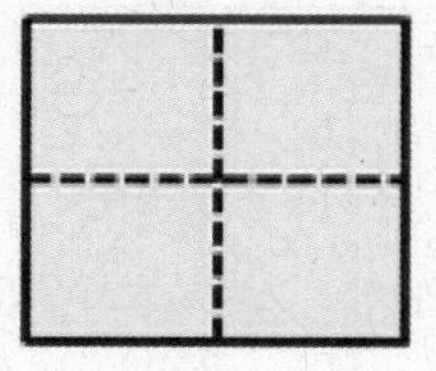

Read **Draw** **Write**

Name ______________________________ Date ______________

1. Label the side lengths of the shaded and unshaded rectangles when needed. Then, find the total area of the large rectangle by adding the areas of the two smaller rectangles.

a.

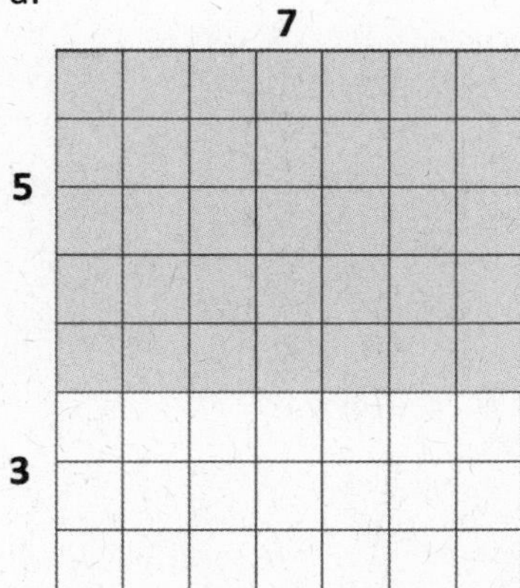

$8 \times 7 = (5 + 3) \times 7$

$= (5 \times 7) + (3 \times 7)$

$=$ ______ $+$ ______

$=$ ______

Area: ______ square units

b.

$12 \times 4 = ($______ $+ 2) \times 4$

$= ($______ $\times 4) + (2 \times 4)$

$=$ ______ $+ 8$

$=$ ______

Area: ______ square units

c.

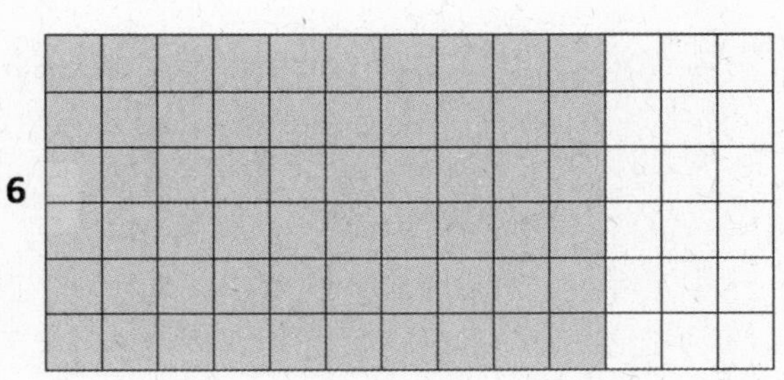

$6 \times 13 = 6 \times ($______ $+ 3)$

$= (6 \times$ ______ $) + (6 \times 3)$

$=$ ______ $+$ ______

$=$ ______

Area: ______ square units

d.

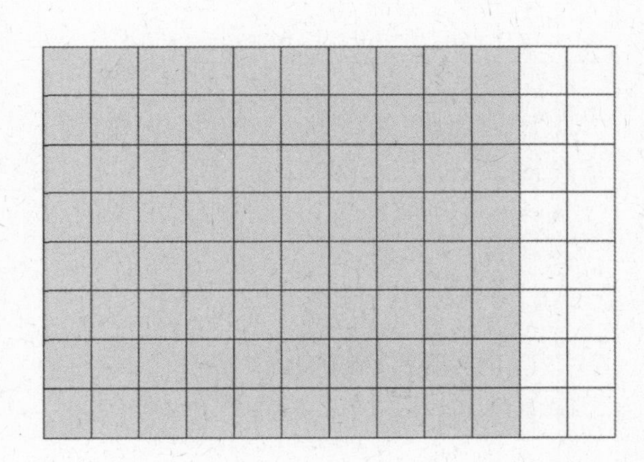

$8 \times 12 = 8 \times ($______ $+$ ______$)$

$= (8 \times$ ______ $) + (8 \times$ ______$)$

$=$ ______ $+$ ______

$=$ ______

Area: ______ square units

2. Vince imagines 1 more row of eight to find the total area of a 9 × 8 rectangle. Explain how this could help him solve 9 × 8.

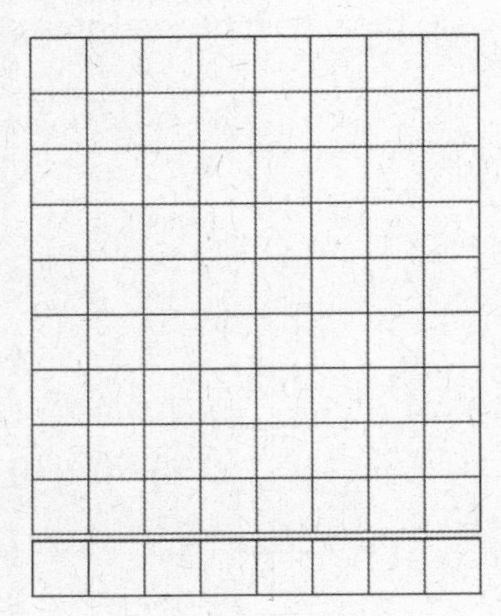

3. Break the 15 × 5 rectangle into 2 rectangles by shading one smaller rectangle within it. Then, find the sum of the areas of the 2 smaller rectangles and show how it relates to the total area. Explain your thinking.

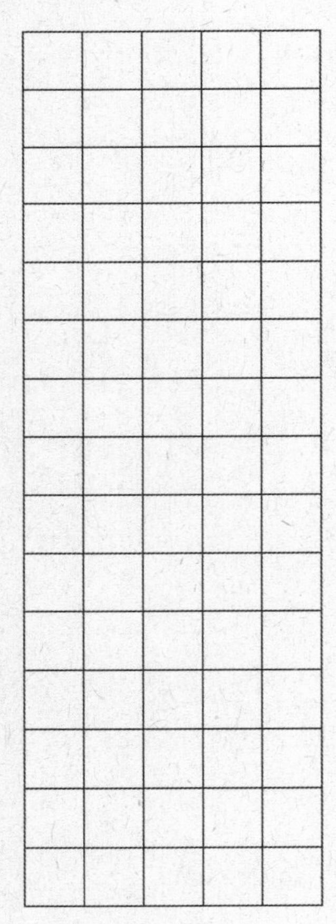

Name ______________________________ Date ______________

Label the side lengths of the shaded and unshaded rectangles. Then, find the total area of the large rectangle by adding the areas of the 2 smaller rectangles.

1.

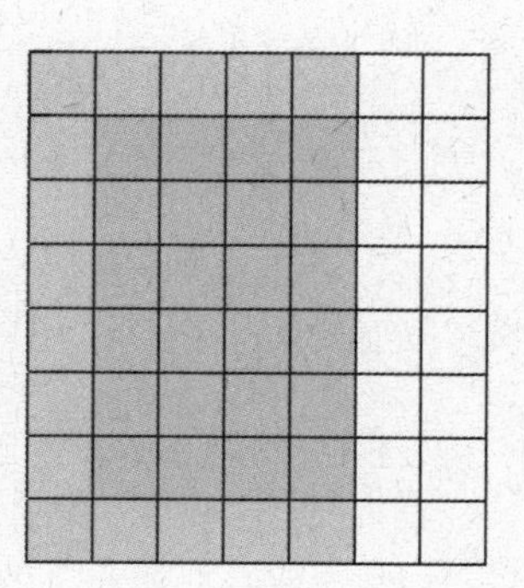

$8 \times 7 = 8 \times (\_\_\_\_ + \_\_\_\_)$

$= (8 \times \_\_\_\_) + (8 \times \_\_\_\_)$

$= \_\_\_\_ + \_\_\_\_$

$= \_\_\_\_$

Area: ______ square units

2.

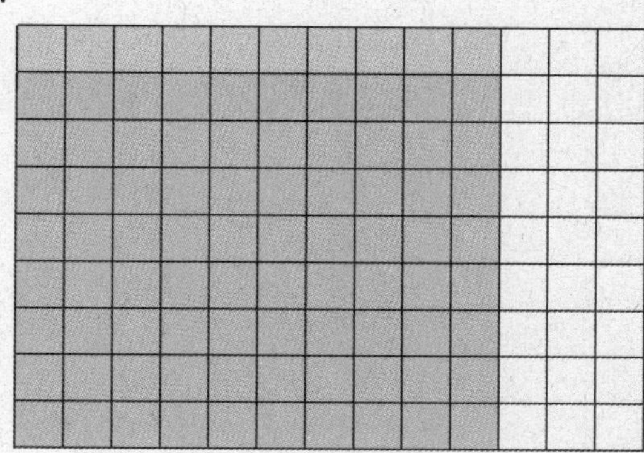

$9 \times 13 = 9 \times (\_\_\_\_ + \_\_\_\_)$

$= (\_\_\_\_ \times \_\_\_\_) + (\_\_\_\_ \times \_\_\_\_)$

$= \_\_\_\_ + \_\_\_\_$

$= \_\_\_\_$

Area: ______ square units

tiling

A table in a restaurant measures 3 feet by 6 feet. For a large party, workers at the restaurant place 2 tables side by side to create 1 long table. Find the area of the new, longer table.

Read **Draw** **Write**

Name ____ Date ____________________

1. The rectangles below have the same area. Move the parentheses to find the unknown side lengths. Then, solve.

a.

6 cm

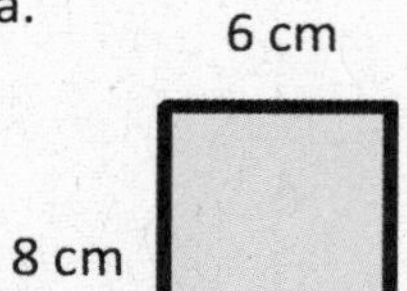

8 cm

Area: $8 \times$ _______ = _______

Area: _______ sq cm

b.

_______ cm

1 cm

Area: $1 \times 48 =$ _______

Area: _______ sq cm

c.

_______ cm

2 cm

Area: $8 \times 6 = (2 \times 4) \times 6$

$= 2 \times 4 \times 6$

= _______ × _______

= _______

Area: _______ sq cm

d.

_______ cm

4 cm

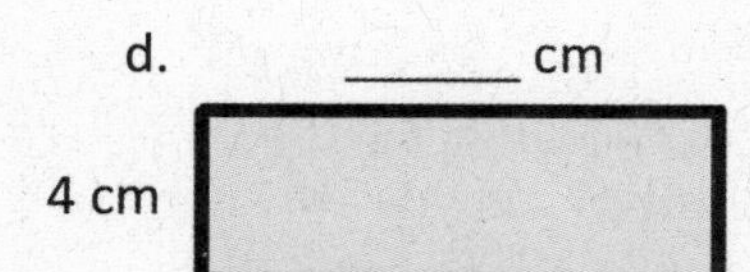

Area: $8 \times 6 = (4 \times 2) \times 6$

$= 4 \times 2 \times 6$

= _______ × _______

= _______

Area: _______ sq cm

e.

_______ cm

_______ cm

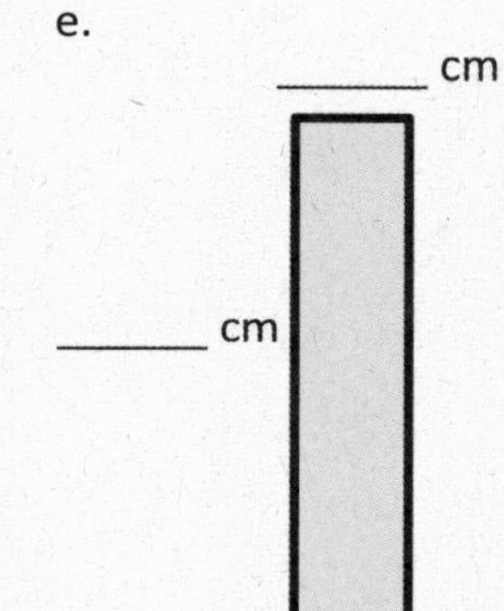

Area: $8 \times 6 = 8 \times (2 \times 3)$

$= 8 \times 2 \times 3$

= _______ × _______

= _______

Area: _______ sq cm

2. Does Problem 1 show all the possible whole number side lengths for a rectangle with an area of 48 square centimeters? How do you know?

3. In Problem 1, what happens to the shape of the rectangle as the difference between the side lengths gets smaller?

4. a. Find the area of the rectangle below.

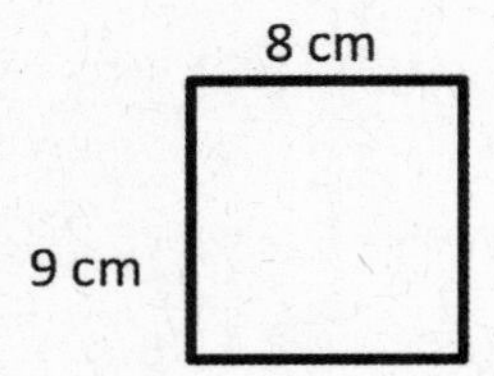

b. Julius says a 4 cm by 18 cm rectangle has the same area as the rectangle in Part (a). Place parentheses in the equation to find the related fact and solve. Is Julius correct? Why or why not?

$4 \times 18 = 4 \times 2 \times 9$

$= 4 \times 2 \times 9$

$=$ ______ $\times$ ______

$=$ ______

Area: ______ sq cm

c. Use the expression 8×9 to find different side lengths for a rectangle that has the same area as the rectangle in Part (a). Show your equations using parentheses. Then, estimate to draw the rectangle and label the side lengths.

Name ______________________________ Date ______________

1. Find the area of the rectangle.

8 cm

8 cm

2. The rectangle below has the same area as the rectangle in Problem 1. Move the parentheses to find the unknown side lengths. Then, solve.

______ cm

______ cm

Area: $8 \times 8 = (4 \times 2) \times 8$

$= 4 \times 2 \times 8$

$=$ ______ $\times$ ______

$=$ ______

Area: _______ sq cm

a. Find the area of a 6 meter by 9 meter rectangle.

b. Use the side lengths, 6 m × 9 m, to find different side lengths for a rectangle that has the same area. Show your equations using parentheses. Then estimate to draw the rectangle and label the side lengths.

Read **Draw** **Write**

Name ______________________________ Date ______________

1. Each side on a sticky note measures 9 centimeters. What is the area of the sticky note?

2. Stacy tiles the rectangle below using her square pattern blocks.

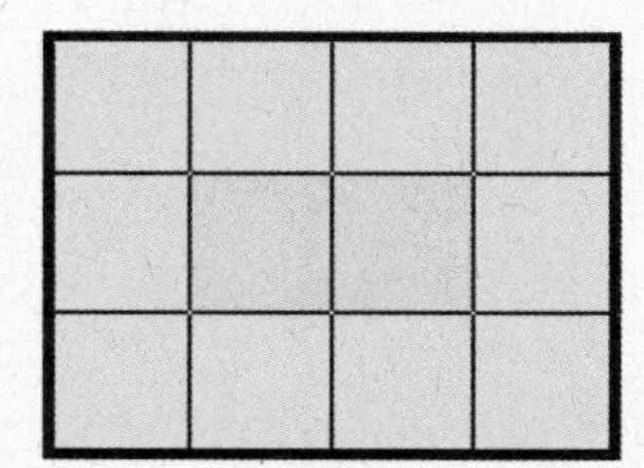

   a. Find the area of Stacy's rectangle in square units. Then, draw and label a different rectangle with whole number side lengths that has the same area.

   b. Can you draw another rectangle with different whole number side lengths and have the same area? Explain how you know.

3. An artist paints a 4 foot × 16 foot mural on a wall. What is the total area of the mural? Use the break apart and distribute strategy.

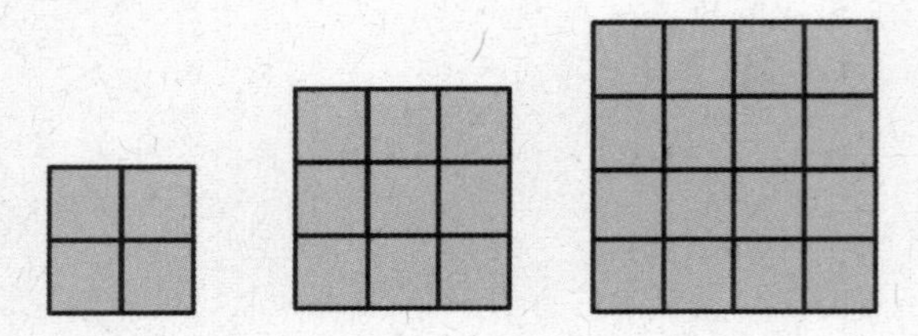

4. Alana tiles the 3 figures below. She says, "I'm making a pattern!"

a. Find the area of Alana's 3 figures and explain her pattern.

b. Draw the next 2 figures in Alana's pattern and find their areas.

5. Jermaine glues 3 identical pieces of paper as shown below and makes a square. Find the unknown side length of 1 piece of paper. Then, find the total area of 2 pieces of paper.

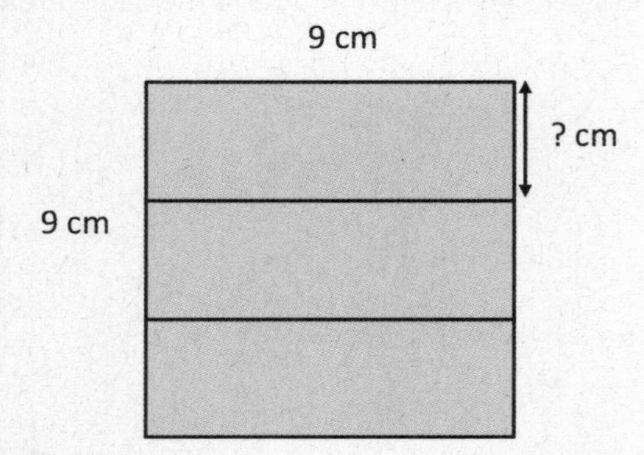

Name ____ Date ____________________

1. A painting has an area of 63 square inches. One side length is 9 inches. What is the other side length?

9 inches

Area = 63 square inches

2. Judy's mini dollhouse has one floor and measures 4 inches by 16 inches. What is the total area of the dollhouse floor?

Anil finds the area of a 5-inch by 17-inch rectangle by breaking it into 2 smaller rectangles. Show one way that he could have solved the problem. What is the area of the rectangle?

Read **Draw** **Write**

Name ______________________________ Date ______________

1. Each of the following figures is made up of 2 rectangles. Find the total area of each figure.

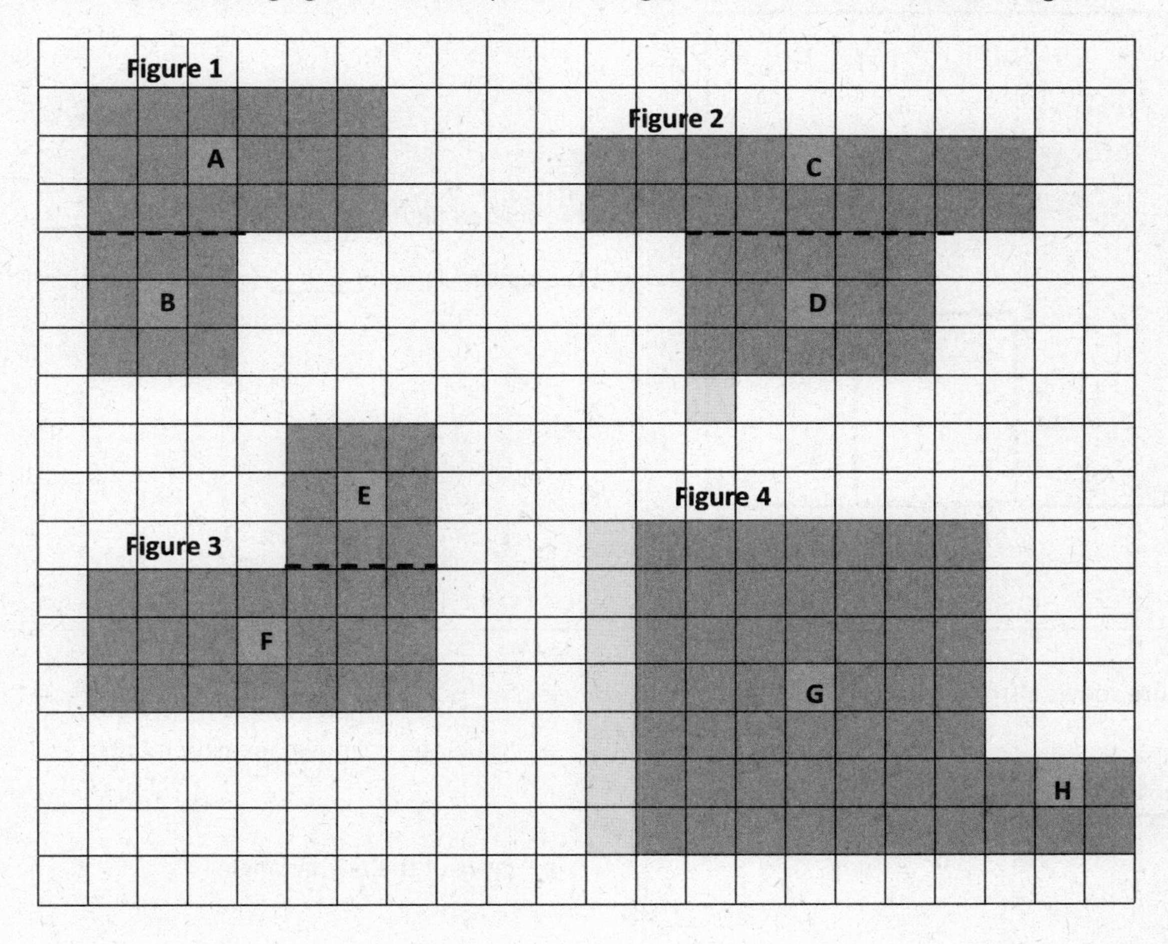

Figure 1: Area of A + Area of B: ___18___ sq units + ________ sq units = ________ sq units

Figure 2: Area of C + Area of D: ________ sq units + ________ sq units = ________ sq units

Figure 3: Area of E + Area of F: ________ sq units + ________ sq units = ________ sq units

Figure 4: Area of G + Area of H: ________ sq units + ________ sq units = ________ sq units

2. The figure shows a small rectangle cut out of a bigger rectangle. Find the area of the shaded figure.

9 cm

10 cm

3 cm

4 cm

Area of the shaded figure: ______ – ______ = ______

Area of the shaded figure: ______ square centimeters

---

3. The figure shows a small rectangle cut out of a big rectangle.

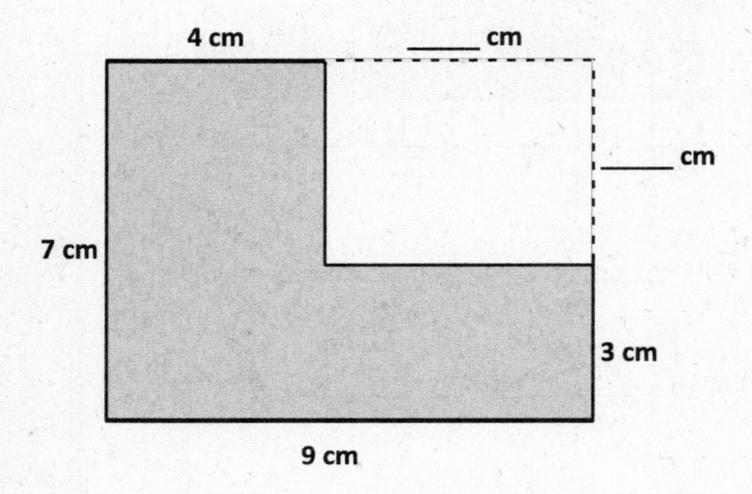

a. Label the unknown measurements.

b. Area of the big rectangle:

_____ cm × _____ cm = _____ sq cm

c. Area of the small rectangle:

_____ cm × _____ cm = _____sq cm

d. Find the area of the shaded figure.

Name ______________________________________ Date ____________________

The following figure is made up of 2 rectangles. Find the total area of the figure.

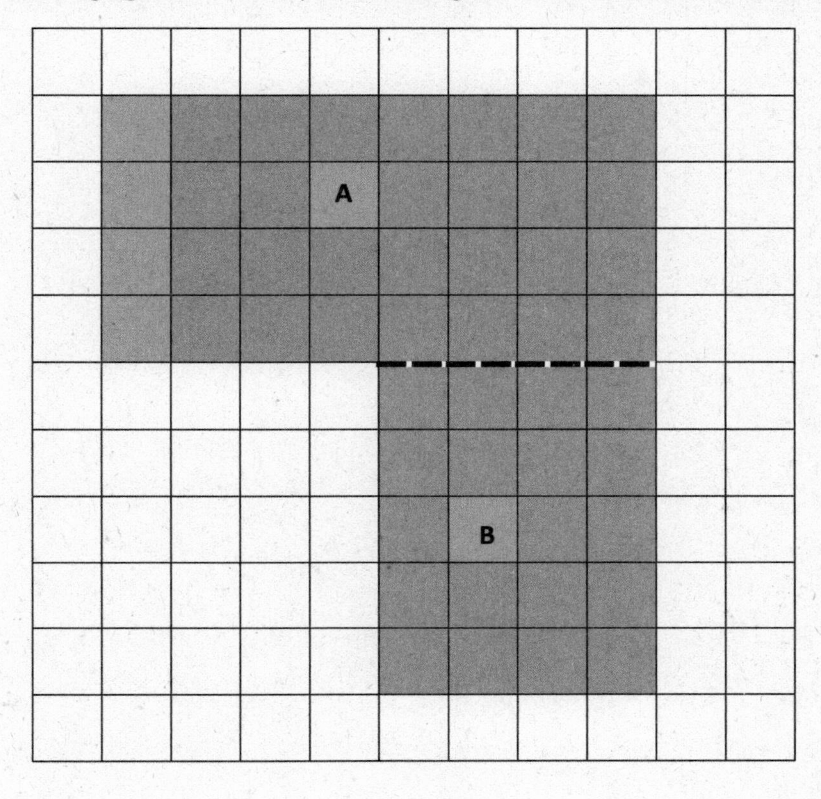

Area of A + Area of B: ________ sq units + ________ sq units = ________ sq units

large grid

a. Break apart the shaded figure into 2 rectangles. Then, add to find the area of the shaded figure below.

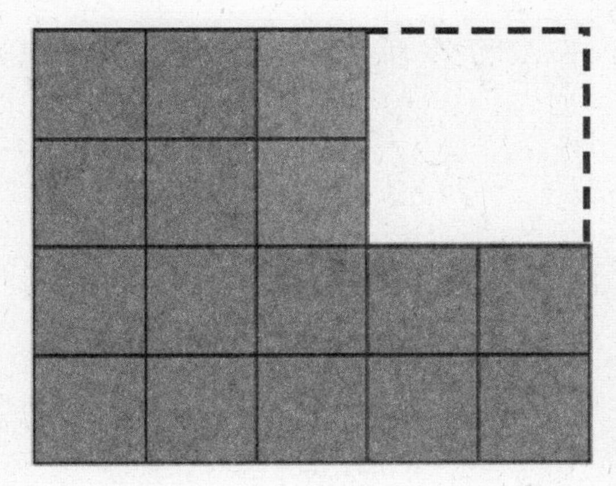

b. Subtract the area of the unshaded rectangle from the area of the large rectangle to check your answer in Part (a).

Read **Draw** **Write**

Name ______________________________ Date ______________

1. Find the area of each of the following figures. All figures are made up of rectangles.

a.

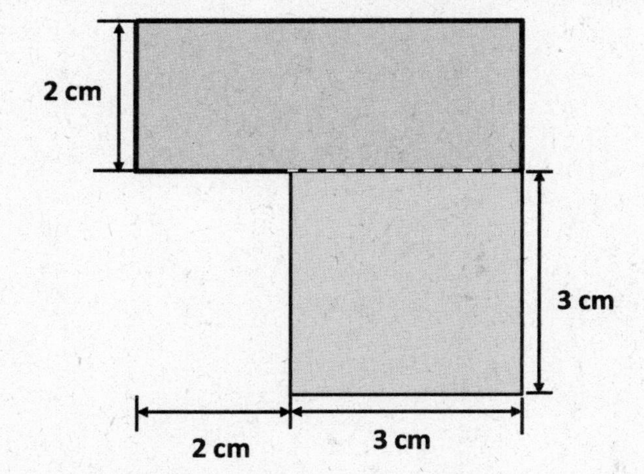

b.

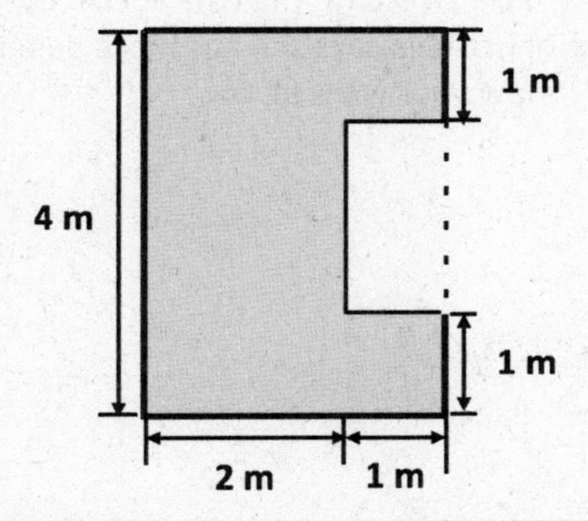

2. The figure below shows a small rectangle in a big rectangle. Find the area of the shaded part of the figure.

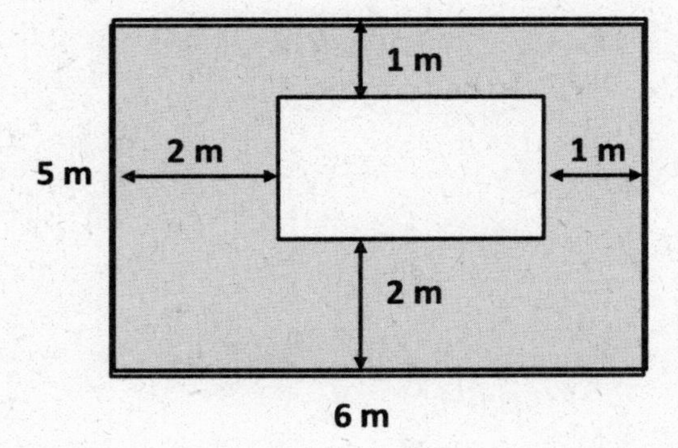

3. A paper rectangle has a length of 6 inches and a width of 8 inches. A square with a side length of 3 inches was cut out of it. What is the area of the remaining paper?

4. Tila and Evan both have paper rectangles measuring 6 cm by 9 cm. Tila cuts a 3 cm by 4 cm rectangle out of hers, and Evan cuts a 2 cm by 6 cm rectangle out of his. Tila says she has more paper left over. Evan says they have the same amount. Who is correct? Show your work below.

Name ______________________________ Date ________________

Mary draws an 8 cm by 6 cm rectangle on her grid paper. She shades a square with a side length of 4 cm inside her rectangle. What area of the rectangle is left unshaded?

Name ______________________________ Date ______________

1. Make a prediction: Which room looks like it has the biggest area?

2. Record the areas and show the strategy you used to find each area.

Room	Area	Strategy
Bedroom 1	_______ sq cm	
Bedroom 2	_______ sq cm	
Kitchen	_______ sq cm	
Hallway	_______ sq cm	
Bathroom	_______ sq cm	
Dining Room	_______ sq cm	
Living Room	_______ sq cm	

3. Which room has the biggest area? Was your prediction right? Why or why not?

4. Find the side lengths of the house without using your ruler to measure them, and explain the process you used.

   Side lengths: __________ centimeters and __________ centimeters

5. What is the area of the whole floor plan? How do you know?

   Area = __________ square centimeters

The rooms in the floor plan below are rectangles or made up of rectangles.

Bedroom 1		Bathroom
Kitchen	Hallway	Bedroom 2
Dining Room	Living Room	

Name ____________________________________ Date ____________________

Jack uses grid paper to create a floor plan of his room. Label the unknown measurements, and find the area of the items listed below.

Desk

Table

Bed

Dresser

Name	Equations	Total Area
a. Jack's Room		________ square units
b. Bed		________ square units
c. Table		________ square units
d. Dresser		________ square units
e. Desk		________ square units

Name ______________________________ Date ________________

Record the new side lengths you have chosen for each of the rooms and show that these side lengths equal the required area. For non-rectangular rooms, record the side lengths and areas of the small rectangles. Then, show how the areas of the small rectangles equal the required area.

Room	New Side Lengths
Bedroom 1: 60 sq cm	
Bedroom 2: 56 sq cm	
Kitchen: 42 sq cm	

Hallway: 24 sq cm	
Bathroom: 25 sq cm	
Dining Room: 28 sq cm	
Living Room: 88 sq cm	

Name ______________________________ Date ______________

Find the area of the shaded figure. Then, draw and label a rectangle with the same area.

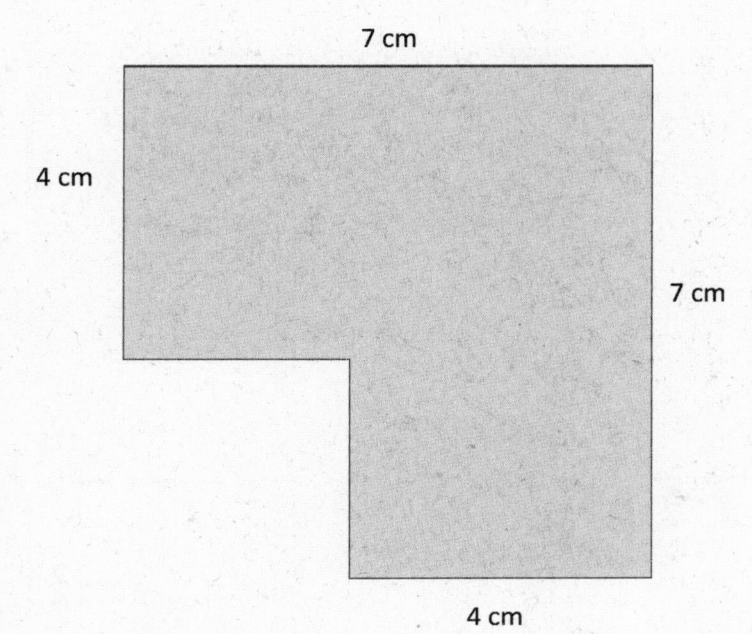

Credits

Great Minds® has made every effort to obtain permission for the reprinting of all copyrighted material. If any owner of copyrighted material is not acknowledged herein, please contact Great Minds for proper acknowledgment in all future editions and reprints of these modules.